T0123167

Subsea Rock Installation and Rock Placement

SUBSEA ROCK INSTALLATION AND ROCK PLACEMENT

A Guide for Offshore Operators

Robert Eadie

MCMI, MCICES, MIMarEST, CMarTech

Whittles Publishing

Published by
Whittles Publishing,
Dunbeath,
Caithness KW6 6EG,
Scotland, UK

www.whittlespublishing.com

ISBN 978-184995-454-9

Contents

About the author

Robert Eadie has over 40 years seagoing experience, with the last 23 years spent in the offshore industry. He has worked almost exclusively as a freelance offshore client representative since late 2005, mainly at sea in the oil and gas sector. In his assignments he has been in all types of vessel from small inshore craft to large construction and rock installation vessels and numerous craft of opportunity. He has been a client representative offshore in over 40 rock installation tours from 2006, on board the fall pipe vessels *Sandpiper*, *Seahorse*, *Tertnes*, *Tideway Rollingstone*, *Simon Stevin*, *Stornes*, *Flintstone*, *Nordnes*, *Rockpiper* and *Bravenes* and the multipurpose vessel *Adhemar de Saint-Venant*. These assignments have ranged from large major FPSO-based oil and gas field development projects to short duration seabed preparation and cable crossing projects. In summer 2003 he oversaw coastal engineering works from the side stone dumper *HAM 601*. His subsea rock installation clients include fifteen oil and gas operators, three offshore construction companies, one utility and one telecom cable company.

He is a member of The Chartered Management Institute, The Chartered Institute of Civil Engineering Surveyors, the Institution of Marine Engineering Science and Technology, The Hydrographic Society and is a Chartered Marine Technologist. He has worked at sea for the vast majority of his working life and at every level; as a seafarer, member of the project team or as a client representative. He served in the Royal Navy from 1974 to 1993. After gaining his bridge watchkeeping and ocean navigation certificates, he specialised as a hydrographic survey officer in 1981. He achieved International Hydrographic Organisation (IHO) Category A Surveyor status in 1986, upon successful completion of the Royal Navy Long Hydrographic Course for Officers. He was the Executive Officer and second in command of the Royal New Zealand Navy ocean survey ship HMNZS *Monowai* from Oct 89 to Jun 91. His final appointment in the Royal Navy was as the Commanding Officer of Naval Party 1008. This was a Royal Navy survey unit based in HM Naval Base Devonport working at sea in a commercial vessel to progress the UK's Civil Hydrographic Programme.

On completion of his Royal Navy service he spent 12 years in the Devon and Cornwall Constabulary as a Special Constable serving mainly in rural Devon including part of Dartmoor, whilst between offshore jobs.

Preface

I was privileged to serve in the Royal Navy for over 19 years and to become a Hydrographic Survey Officer. Royal Navy survey ships and units are tasked with carrying out nautical charting and defence related surveys. As a mariner and user of nautical charts, I understood the information needed for the safe navigation of a ship. Despite having transferrable skills and a baseline of knowledge, I quickly learned that there were a number of users of survey data throughout an offshore project, all seeking different aspects at various stages.

When I was approached to write this book, I was taken aback. I had taken knowledge of rock installation for granted and laboured under the impression that it was understood by many users. By looking critically at SRI operations offshore and engaging in discussion with project staff onshore, I observed that there is a level of anxiety concerning rock installation. This anxiety is understandable and no doubt stems from unfamiliarity, combined with the costs involved in a bespoke product. Project staff can be reassured that the rock installation contractors are experts in their field. The purpose of this book is to allay that anxiety and to assist users in beginning to understand this important engineering tool. My intention is to enable rock installation service users to start their dialogue with their contractors from an informed position.

It is based upon my personal experience as a client representative offshore in a series of rock installation projects from 2003 onwards. I have worked mainly in the oil and gas sector on a broad range of projects with some coastal engineering, power and telecom cable projects to add to my knowledge. It draws on my experience gained in the UK, Norwegian and Danish Exclusive Economic Zones, where rock installation can be an essential element in a significant number of projects. It also benefits from my understanding of large oil field development projects to those of a much more specialised nature; such as the preparation of sites from which a jackup rig can operate safely.

The book sets the scene and places rock installation in the context of the offshore industry today. It is a mature industry where health, safety and environmental considerations are factored into a project with the same level of professionalism as regulatory compliance and good design. In this volume, the reader will be introduced or refamiliarised with the planning, preparation, execution and reporting of a rock installation project.

Any volume about rock installation would be incomplete without some discussion of seabed soils, rock, the basic material, and the relationship between the quarry and the receiving vessel, which together form a system of turning a geological formation onshore into a rock berm offshore. Rock installation vessels, their characteristics, equipment and capabilities are examined in outline.

The marine environment is also discussed. I am firmly of the opinion that all personnel working in the offshore industry, whether at sea or in onshore positions need to understand the harsh but majestic, hostile and often unforgiving marine environment. Furthermore, offshore industry professionals should have an appreciation and awareness of both the beneficial and the disadvantageous effects of human activity upon the marine environment and the creatures that inhabit it.

With the cyclic nature of the offshore oil and gas industry and a number of new operators developing both end of life and small new fields, knowledge concerning SRI has been diluted or has even evaporated. I hope that this book is both a useful and interesting enabler for project staff and offshore client representatives who are new to rock installation such that they have a fulfilling rock installation experience, from the initial discussions with a contractor to the final results and verification.

Robert Eadie

Acknowledgements

The author would like to acknowledge the help received from:

Mr Donald Reid for his foreword and proof reading of the final draft;

Captains Wim Deca, Bjorn Van Reit and Frederick Bruyninckx of DPFPV *Simon Stevin*;

Mr Tom Quintelier Offshore Works Manager DPFPV *Simon Stevin*;

Mr Scott Adams for checking operational technical details;

Mr Mitch Foster for permitting the use of his photograph of side chute SRI;

Mr Rob Harding for checking geotechnical information;

Mr Trevor Horne a former Royal Navy colleague and author for his guidance on writing;

Mr Gordon Stewart for checking details of HSE and renewables;

Mr Barry van der Meijden for permitting the use of his photograph of *Bravenes*;

Enquest Britain Limited and Mr George Tulloch for allowing me to use an as-built plan;

Jan De Nul Group Public Relations and Rock Installation Departments and

Van Oord Public Relations for the provision of photographs used in the book.

He also thanks all personnel in fall pipe vessels he has sailed in, who have always shown their professionalism, enthusiasm, proficiency, teamwork, patience and engaging openness.

Disclaimer

This is a guide to assist project personnel involved in a rock installation task. It is based on the author's experience of rock installation and his wider experience as a mariner from 1974 onwards. It will enable the reader to gain some insight into subsea rock installation and enter into an informed dialogue with the contractor with an understanding of what can and, more importantly, what cannot be done.

It cannot take any precedence over national or international regulation or convention, client company or contractor's procedures, method statements and other protocols.

Unless otherwise stated all photographs are © Robert Eadie. The author has asked companies for permission to use images of their operations and appropriate acknowledgements are included.

A highly detailed technical reference covering the uses of rock in offshore engineering projects can be found in *The Rock Manual - The use of rock in hydraulic engineering*, ISBN 978-0-86017-683-1 Edition 2 (2007).

Foreword

The author and I started our seagoing careers in the seventies in the Royal Navy and have both gravitated to oil and gas offshore operations. Delivering rock to the seabed in a controlled and precise manner as part of a specialist team came later! I have had the pleasure to serve on a rock placement vessel with the author and several of those also kindly acknowledged.

My first experience with rock placement or installation was a surprise. In those days it was rather disparagingly called rock dumping and I thought it was going to be a fairly coarse and simple process. That could not have been further from the truth. What I discovered very quickly was that a close knit team of professionals had developed great teamwork and applied science and engineering to achieve incredible results. How I wish I had a publication like this before my first trip or as a reference document thereafter.

Donald Reid, AFNI, aIOSH

Abbreviations

BOP	bottom of pipe
CCTV	closed circuit television
CTV	crew transfer vessel (wind farms)
DARO	day rate operations
DEPCON	Deposit Consent UK
DGPS	differential global positioning system (precise navigation using GPS)
DOC	depth of cover
DOL	depth of lowering
DP	dynamic positioning
DPA	designated person ashore
DPO	dynamic positioning operator
DPR	daily progress report
DSV	diving support vessel
DTM	digital terrain model
EEZ	Exclusive Economic Zone – up to 200 miles from the coast or baseline or mutually agreed median between two nations e.g. UK and Norway in the Northern North Sea
EPIC	engineering, procurement, installation and commissioning
FFPV	flexible fall pipe vessel
FJ	filed joint
FPROV	fall pipe remotely operated vehicle
FPROT	fall pipe remotely operated tool (new term for FPROV)
FPSO	floating production and storage offshore
FPV	fall pipe vessel
FRC	fast rescue craft
ftp	file transfer protocol
GIS	geographic information system
GNSS	global navigation satellite system

GPS	global positioning system
GWO	Global Wind Organisation
HF	high frequency (radio frequencies between 3 and 30 MHz)
HiPAP	high precision acoustic positioning
HIRA	hazard identification risk assessment
HPR	hydroacoustic positioning reference
HSE	Health Safety and Environment and Health and Safety Executive (UK)
IMO	International Maritime Organisation
ILO	International Labour Organisation
IMCA	International Marine Contractor's Association
ISPS	International Ship and Port Facility Security
JV	joint venture
KP	kilometre post – indicating distance along a pipeline; the convention is increasing KP in the direction of flow
LAT	lowest astronomic tide
LOLER	Lifting Operations and Lifting Equipment Regulations 1998 (UK)
MARPOL	marine pollution
MBES	multibeam echo sounder
MCA	Marine and Coastguard Agency UK
MIST	minimum industry safety training
MMO	Marine Management Organisation (UK) also marine mammal observer
MOC	management of change
MOD(U)	mobile drilling unit
MSL	mean sea level
ObsROV	observation remotely operated vehicle
OGA	UK Oil and Gas Authority
OGUK	Oil and Gas UK
OIM	offshore installation manager
OOS	out of straightness
OSS	offshore substation
PLET	pipeline end termination
PPE	personal protective equipment
PSV	platform support vessel
PUWER	Provision and Use of Work Equipment Regulations 1998 (UK)
RIDDOR	Reporting of Injuries, Diseases and Dangerous Occurrences Regulations 2013 (UK)
ROV	remotely operated vehicle

ROVSV	ROV equipped survey vessel
SIMOPS	simultaneous operations
SOLAS	Safety of Life at Sea
SRI	subsea rock installation
Te	metric tonne (1000 kg)
TOFS	time out for safety
TOP	top of pipe
UHB	upheaval buckling
UNCLOS	United Nations Convention on the Law of the Sea
USBL	ultra short baseline
VHF	very high frequency (radio frequencies between 30 and 300 MHz)
WROV	work class remotely operated vehicle

Glossary of terms

500 m safety zone	Designated area around a platform or floating asset, where the platform or asset OIM controls access and activity. When working in a safety zone vessel masters have additional responsibilities and are obliged to report any incidents to the platform. Unauthorised entry is not permitted.
beam	Width of a ship.
counter fill	Rock structure upon which a main berm will be installed to provide support, protection or stabilisation.
critical bend radius	The minimum radius for a product.
depth of cover	Vertical distance between the top of the product and top of cover.
depth of lowering	Vertical distance between the mean undisturbed seabed and the top of the product
depth of trench	Vertical distance between the mean undisturbed seabed and the bottom of the trench
draught	The depth of a ship in water measured from the keel.
Exclusive Economic Zone	An area beyond a coastal state's 12 mile territorial waters, that can reach up to 200 miles or to a mutually agreed median line between adjacent coastal states, e.g. the UK has median lines with Ireland, Faroes, Norway, Germany, Denmark, Netherlands, Belgium and France.
free span	Unsupported length of pipeline or cable. The free span may be caused by a combination of seabed topography and pipe lengths and rigidity or by scouring action of tidal stream or current or by fish and crustacea. Some free span lengths

	are acceptable, whilst others have to be rectified depending upon the pipeline or cable.
hove to	Vessel is making little or no headway with ship's bows into a heavy sea
lowest astronomic tide (LAT)	The lowest tide level, which can be predicted to occur under average meteorological conditions and under any combination of astronomical conditions.
mean sea level (MSL)	Mean sea level when measured relative to a land datum. This is a variable level as more data is gathered from a larger number of tide gauges.
mean undisturbed seabed	The mean value of the nearest seabed unaffected by trenching or ploughing activity to the left and right of the trench
penetration	Rock loss through passing into the seabed soil, especially relevant in a soft seabed. The term sinkage is also used.
product	Generic term used to describe, pipelines, umbilicals, jumpers and cables
rock berm	Rock structure placed on the seabed.
rock grade	The rock grade can be quoted as a range in either metric or imperial measurement to cover the size of rock to be used. The industry standard is 1 inch to 5 inch or 25 mm to 122 mm. Other grades are available, often to order.
rock placement	The precise and controlled positioning of rock on the seabed to build rock berms, interchangeable with subsea rock installation.
scope of work	A summary of the work to be done; the term workscope may be used.
sinkage	Rock loss through passing into the seabed soil, especially relevant in a soft seabed. The term penetration is also used.
subsea rock installation	The precise and controlled positioning of rock on the seabed to build rock berms, interchangeable with rock placement
theoretical tonnage	The calculated volume of rock multiplied by the dry bulk weight, which is typically 1.55 Te m^3
vessel heave	Linear vessel movement in the vertical plane.
vessel pitch	Angular vessel movement forward to aft in the vertical plane.
vessel toll	Angular vessel movement from side to side in the vertical plane.

vessel surge	Linear vessel movement forward and aft
vessel sway	Linear vessel movement to port and starboard
vessel yaw	Angular vessel movement from bow to stern in the horizontal plane.
workscope	A summary of the work to be done; the term scope of work may be used.

1 Introduction

This short guide is primarily intended for users of, and those who procure, subsea rock installation (SRI) services. This group includes but is not restricted to: project managers, project engineers, project surveyors and offshore representatives working on coastal and civil engineering, oil and gas, renewable energy, power and telecom cable projects with an offshore or marine element.

Subsea rock installation is one of many phases required for the completion of a successful offshore project. This book covers the breadth of subsea rock installation including HSE, purposes, planning, methods, operations, logistics, quality control and verification. Regulation regarding shipping and seafarers is also outlined. This guide draws on the author's experiences in the offshore industry in general and in particular of overseeing subsea rock installation at sea.

A rock installation phase may require in excess of a million tonnes of rock placed over a number of campaigns over some years. Alternatively, it may be a small proportion of a vessel load placed in a few hours. SRI projects vary in purpose and complexity. The highly specialised vessels, organisation, methods and finished deliverables are discussed in later chapters.

Subsea rock installation has broadened in scope from its embryonic beginnings in coastal engineering and land reclamation projects in the Netherlands in the 1950s. As needs arose in the offshore oil and gas industry, rock installation was embraced as a solution. In 2006 there were five operational rock installation vessels operated by three companies with an aggregate load capacity of around 78,000 Te. At the time of writing there are now eleven active rock installation vessels operated by four companies with an aggregate load capacity approaching 200,000 Te. One vessel is currently being converted to a fall pipe vessel from a heavy lift ship and is expected to have a capacity of around 17,000 Te. Other conversions or new builds may follow to replace older vessels nearing the end of their service and to meet the needs of a growing and diversifying market.

Rock dumping

The term rock dumping, which is a direct translation into English from Dutch, was used to describe the use of rock in land reclamation and coastal engineering projects from small specialist coastal vessels and barges. It was then adopted by the offshore oil and gas industry to include work carried out by large fall pipe vessels, equipped with an array of above water and subsea precise navigation and control systems.

As rock installation techniques and vessels evolved and uses for rock installation expanded, the term rock dumping became more widely used and awareness of it increased in the offshore oil and gas industry. However the acceptance of rock dumping as a description of activity did not readily transfer to the renewable energy sector.

Following a misunderstanding with the German authorities who sought clarification on the activity, the terms subsea rock installation or rock placement were adopted. These are terms that more accurately describe the precise and carefully controlled nature of the operation. Rock dumping was considered to give the impression that it was an indiscriminate disposal of material onto the seabed, which is not and never has been the case. Whilst the phrase rock dumping may still be used throughout the offshore industry, it will not be used any more in this book.

Although the term has not entirely disappeared, derivatives revealing its former prominence are still in use. The detailed work planning onboard forms a dump plan, the software system used to assist in the work is referred to as dump tool and the rate of deposition through the fall pipe, quoted in tonnes per hour, is called the dump rate!

Subsea rock installation is a multidisciplinary activity. Both offshore and onshore personnel will benefit from an appreciation of geology, chemistry, mechanics, engineering, hydrographic surveying and geographic information systems (GIS), oceanography, meteorology, logistics and marine operations.

The author makes no apologies for emphasising that the marine environment is a harsh one; weather and sea conditions together with working close to offshore structures and the need to manage a number of vessels in a small area, all make for a challenging work place. The sea is unforgiving and it deserves attention and respect.

The importance of a well informed and healthy dialogue in the planning stage, combined with highly skilled and qualified personnel cannot be taken for granted. The high standards extend offshore, where the contractors provide well maintained, reliable vessels and equipment.

The offshore oil and gas industry had its origins in the United States of America. Many conventions still use imperial measurement in specification documents. Pipeline sizes tend to be quoted in inches, a standard pipeline section is 40 feet long (12.5 m). Rock sizes are often quoted in inches, although metric units are becoming more widespread.

Subsea rock installation is a bespoke product, which carries a premium. However, it can rarely, if ever, be viewed as a risk to a project. Apart from the usual possibility of delays caused by schedule overruns, hostile environmental conditions and the occasional defect, the time a rock installation vessel spends on task can be accurately forecast, when the scope of work is specified. Rock placement provides a flexible, versatile and essential option for a range of offshore projects where product protection, support and stabilisation are required. The contractors carrying out subsea rock installation are experts in their field, backed up by experienced project and engineering support staff onshore.

Delft University in the Netherlands, working in co-operation with the rock installation contractors, is a leading body for the theory, modelling, development and testing of SRI techniques. The guide is set out to take the reader from the initial planning stage where options are being considered through detailed planning to operations offshore and the final deliverables. It will introduce the intricacies of the work and the skilled teamwork needed to carry out the task safely and proficiently.

DPFPV *Seahorse* from DSV *Toisa Polaris*, Skarv field June 2010

Jelsa Quarry near Stavanger

DPFFPV *Stornes* departing Kristiansund with a load

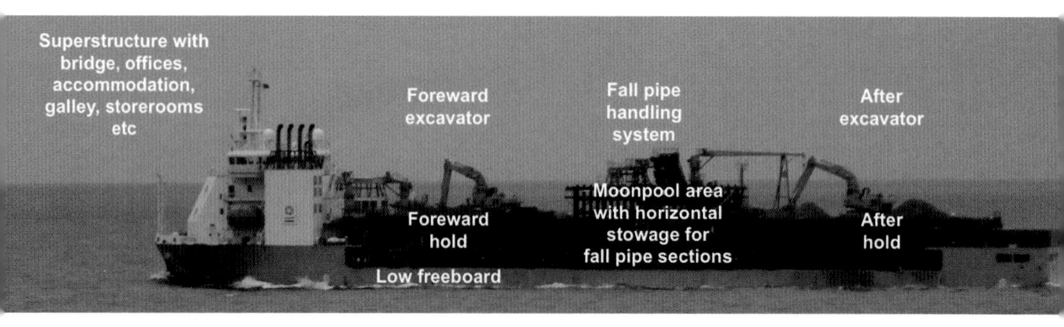

Flintstone loaded June 2013

FFPV *Stornes* outbound left and FFPV *Nordnes* inbound pass off Kristiansund September 2015

The author embraces the belief that health, safety, environmental consideration, quality control and quality assurance are integral to the success of a project. These important topics are discussed in outline in the first chapters of the guide.

Any opinions expressed in this guide are those of the author based upon his experience and from his perspective as a client representative offshore, working for both operators and construction contractors, reporting to project staff onshore.

2 Health, safety and environmental

This guide is not intended to be a health and safety manual. However, it should always be borne in mind that health, safety and environmental (HSE) is an integral part of any project. Any guide to any offshore activity would be incomplete without discussing it. HSE should be an overarching consideration in the planning and execution of all projects offshore.

In the UK, offshore operations are covered by the Health and Safety at Work Act 1974. The UK based offshore oil and gas industry is regulated by the Health and Safety Executive (HSE) Energy Division. UK offshore renewable energy sites are also included in the scope of HSE legislation and regulation.

The author has witnessed a significant and improved change in health and safety in the offshore industry. It has changed from education and training covering safety from a low baseline which was required in the mid 1990s. Today HSE is focussed upon the rigorous maintenance of a safety culture, its re-enforcement and development of higher standards. This has been achieved by wise investment, the encouragement of open discussion, embracing a no-blame culture and giving ownership of safety matters to personnel in the workplace.

All personnel have a duty to stop the job and ask for TOFS (time out for safety) if they see an unsafe situation arising, unsafe practice or are in any doubt about what is required. The purpose of a TOFS is to make the situation safe and to give personnel time to revisit the risk assessment and toolbox talk (a toolbox talk is a discussion between a supervisor and personnel doing the job which covers the what, how and safety aspects of the task, and participants are encouraged to contribute). It may be necessary to clarify procedures or even make a management of change (MOC) to ensure that the task is completed safely.

Safety noticeboards and safety posters in changing rooms and other areas are very useful. A good poster needs an impact and a precise message.

Vessel master

The vessel master or captain is an experienced mariner with formal training resulting in qualification and certification to demonstrate a level of training and competence. All ship's officers have to carry out regular revalidation as required by national maritime authorities.

The master has the overall responsibility for the safety of the ship and all personnel onboard. The master is also responsible for compliance with maritime law and regulation. His or her experience as a mariner with an instinctive knowledge of the sea adds an unquantifiable value to any project.

Safety reporting

The ease of reporting HSE issues using a safety observation system is without doubt a useful management tool. It can be used to monitor behavioural trends and enable rectification before bad practice leads to personal injury or damage to equipment. The downside is that the safety observation system can be used to report defects and domestic matters that should really be managed elsewhere. On the other hand that can be thought of as worthwhile collateral to engage personnel to get into the habit of raising safety observations. The deck and project teams, who are in the higher profile of activity, can be easily encouraged to raise safety observations. Other teams whose efforts are of equal significance and importance tend to omit safety observation reporting as part of their daily routine. Positive encouragement works.

The offshore oil and gas industry is often perceived as being uncaring with regard to the environment. This is a general misconception and is an ill-informed injustice to the industry, which is highly regulated regarding any damage or threat to the environment.

Hydraulic and other oil spills, no matter how small or accidental, must all be reported via the client-specified management system. For example, in the UK oil loss to the environment is reported via form PON1 within strictly defined time limits. If the vessel is operating within a 500 m safety zone the OIM of the controlling platform or unit must be informed immediately. Vessel masters and superintendents are fully familiar with MARPOL (marine pollution) regulations and the necessary first aid and reporting actions.

HSE planning

No job is ever so important that time cannot be taken to plan and do it safely. Subsea rock installation operations are carried out with all aspects of HSE being a prime consideration. HSE is designed into a project from the outset and not as an afterthought. From the safety case to toolbox talks via HIRA (hazard identification risk assessment) and Level 2 risk assessment, all involved are mindful of safety and the requirement to do no harm to the environment.

The HIRA is a multistage high level risk assessment held by project staff in the office; ideally there should be input from vessel personnel, if they are available. It will cover generic and specific risks associated with the vessel activity and with the project requirements. The final HIRA, which is conducted at the worksite, forms part of the project documentation and participants are encouraged to add input with the recent experience of carrying out the task.

The participants will carry out a hazard identification (HAZID) and identify the associated level of risk. The team will then identify the control measures that must be in place to bring the level of risk to an acceptable level.

Language

English tends to be specified as the project language in many cases. Before considering any technicality, it must be appreciated that most crews are multinational. All managers must be mindful that for many personnel English is a second or even one of several languages they speak. Although it has become a universal language in many areas of the offshore industry, a lack of English can provide many unforeseen chances for misunderstanding and misinterpretation. It is difficult to evaluate the extent of an individual's English knowledge which is derived from their education, cultural background and exposure to English via the media in varying proportions. An assumption of knowledge may compound to give rise to misunderstanding.

Safety notice at Aquarock Quarry, Sandnessjoen

It is therefore important that written and verbal communications are clear and unambiguous and provide the scope for any amplification and an opportunity for questions if required. There is a consensus that the possibility of misunderstanding arising from using a second or third language should be discussed as a risk at the HIRA. It should then be cascaded further to level 2 risk assessment and to tool box talks. The advantage is that it can be discussed in depth with both the supervisors and personnel who are tasked with the work.

HSE at quarries

Quarrying and mining, usually grouped together, are in the unenviable position of being amongst the most hazardous industries on earth with a comparatively high per capita work force fatality and injury rate.

The industry works hard to eliminate hazards and to reduce risk, however the very nature of the quarry environment is unforgiving. Visitors and crew must be briefed over the hazards of working in a quarry. This must include requirements for leaving and joining the vessel, access to the deck area and PPE. Despite attending a vessel safety induction, the author has witnessed a visiting contractor try to leave the ship arbitrarily without any PPE. An alert crewman stopped the job and the contractor was firmly advised on quarry safety by the vessel's master. The contractor was at a very real risk of serious injury from falling rock and moving quarry machinery

Quarry machinery at work

Visiting quarries to load cargo is a day-to-day event for a rock installation vessel. Quarries are potentially very dangerous work places; where there are plenty of hazards including:

- moving heavy machinery
- ultraheavy vehicles
- noise
- artificial light and shadows combining to give poor visibility at night
- loose rock
- rough ground
- unstable surfaces
- airborne rock dust and grit
- mud creating slippery surfaces
- standing water
- poor access to the vessel
- controlled use of explosives

Rock dust

Rock dust can be inhaled and the effects of long term exposure to various types of rock dust are not fully evaluated for all minerals. The hazards arising from exposure to silica, sulphur and mineral asbestos, are very well known. In some quarries, rock can be dampened on the loading belts to reduce the quantity of airborne dust. Vessel air intake filters should be maintained to trap dust before it is distributed throughout the ship. Some dust will enter the vessel through access areas and on footwear and clothing. If excavators are in use whilst loading, the cabin air filter must be capable of trapping fine particles and the operator should have a face mask as optional PPE.

To mitigate its many serious hazards the quarrying industry is heavily regulated in most countries. In practice, vessel and quarry staff have to maintain good effective communication throughout the berthing, loading and departure operations. Personnel and vehicular access, including vessel tenders, to and from the vessel must be coordinated and carefully controlled. At many quarry locations it is impractical to separate foot and vehicle access to the wharf area and extra vigilance is necessary. Personnel waiting for access to and from the vessel should use the appropriate PPE and wait in a designated safe area until instructed by ship's staff or quarry staff as appropriate.

When loaded, the rock in the hold is trimmed to provide a stable load and to prevent rock from moving either by gravity or vessel motion to create a potential hazard to ship stability.

Night loading in winter with rock dust picked up by deck lights

Trimming the load (Reproduced with kind permission of Jan De Nul Group)

Rock installation vessels are required to operate in some of the world's harshest marine environments and occasionally in close proximity to platforms, floating assets or within anchor patterns. Risk assessments should always be treated as dynamic living documents and revisited in the context of the prevailing or anticipated circumstances. For example, the risks of working in low light, darkness and cold need to be highlighted and mitigated when operating in Northern Norway in winter. On the other hand, in tropical areas and during summer, control measures must be in place to prevent the risks of dehydration and exposure to strong sunlight.

Safety at sea

The ship's organisation is structured to make a safe working environment. Bridge staff administer the ship's internal work permit system. All personnel will undergo a vessel induction upon joining and further departmental inductions. An induction is valid for six months and project staff returning after that time must attend another induction. Day visitors are given a shorter safety induction.

When at sea, in addition to the hazards associated with operating in the marine environment, there are further hazards to mitigate. These hazards include natural and man-made subsea obstructions such as shipwrecks, foul ground, project assets and seabed topographic features. It is essential to designate a safe point at which to build up and recover the fall pipe, which must be clear of subsea assets and outside of a 500 m safety zone. There must be a safe route from the fall pipe build up point to and from the work site. This can also be used as a safe escape route to recover from a vessel, platform or drilling unit emergency. It has to be surveyed by the rock installation vessel to ensure that there is a safe fall pipe clearance above the seabed and seabed structures. When working inside an anchor pattern the survey navigation screen must be marked with a safe clearance from the anchor cables, risers and any wet stored and buoyant assets. Locations of any no-rock zones must be marked on the navigation screen.

The ship's staff have always to bear in mind that the hazards and risks, once the fall pipe is deployed, are in three large dimensions. The ship's draught, which is usually small in comparison with length and beam, can become the largest vessel dimension when the fall pipe or FPROV is deployed. With the exception of vessels working in shallow water this is an unusual circumstance for most mariners. The largest rock installation vessels; *Simon Stevin* and *Joseph Plateau* are both 198 m long; the fall pipe depth will often exceed the length of those vessels.

Rock can be successfully placed at depths in excess of 1000 m. At the time of writing Van Oord have successfully completed rock berms at depths in excess of 1200 m. From looking at operator's websites, maximum operating depths of 2000 m are quoted for a number of rock installation vessels. Rock installation at fields on the continental slope is often carried out at depths in excess of 200 m.

No-rock zones may be designated by the geographic limits of any licensing, the presence of subsea structures such as wet stored risers and the jumper or spool side of PLET (pipeline end termination) covers. No-rock zones can also be required to avoid burying rigging, acoustic beacons or placing rock in areas which need to be kept clear for other reasons.

Deck access and PPE

The deck of a rock installation vessel is a heavy industrial site. A such there is a range of potential hazards including moving machinery such as excavators, conveyor belts and fall pipe handling systems, loose rock, wet and possibly muddy slippery decks, machinery noise, and the moon pool area. The moon pool area should be fenced off and personnel whose duties require them to be close to it must use a fall arrester or safety harness.

Access to the deck, when working, is controlled by the bridge. Visitors and other personnel who are unfamiliar with the deck must accompany a ship's

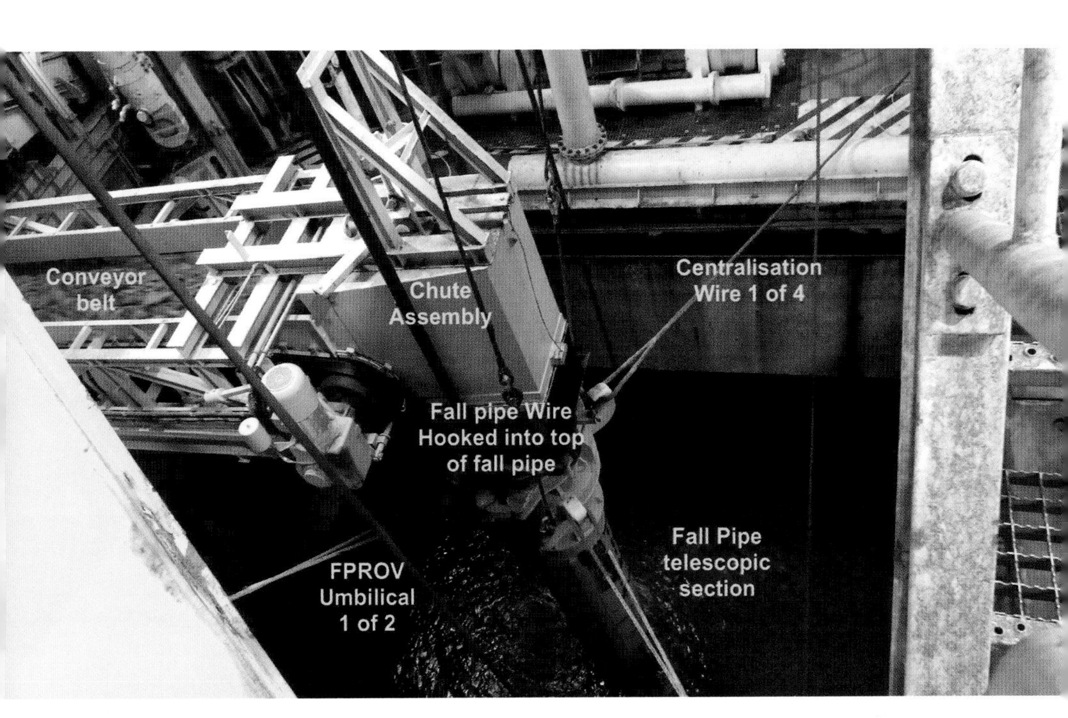

Moon pool area in a fall pipe vessel (Reproduced with kind permission of Jan De Nul Group)

officer who is in radio contact with the bridge when doing a familiarisation, walk around or safety tour. Access to the deck may also be restricted or prohibited altogether in bad weather.

Additional PPE such as ear defenders and eye protection should be used whilst on deck. It is prudent, if not always compulsory. The moon pool area and deck can be very noisy indeed, so ear protection is a must for both visitors and for those personnel who work in the vicinity.

The hazards of dust and grit, both windblown and arising from rock handling, needs to be fully assessed and mitigation put in place. The use of safety glasses and dust masks can mitigate risks arising from exposure to rock dust. Many client companies and most contractors now specify eye protection as a compulsory item of PPE on deck. A recent offshore industry-wide high-profile safety campaign has emphasised the importance of eye protection, which should become habitual for all personnel.

Set up

As well as mitigating the physical risks of operating at sea, the risk of system errors and set ups should be kept in mind. Errors in inputting and checking geodetic and other system details can lead to the misplacement of rock, which not only is a waste of rock, but it may also severely impact project timings.

Small lengths of rock berm have been overlooked in the past. Such an oversight leads to the necessity for remedial works and the additional vessel time and cost involved. It is therefore of paramount importance that the work scope is unambiguous and thoroughly checked to ensure that it has all been captured on the displays and dump plan.

For most GNSS positioning purposes, the raw data is received in the WGS 84 datum and is then transformed into the user's datum. As a cautionary tale, the author has witnessed a satisfactory position check, where the target was positively identified from MBES (multi-beam echo sounder). The operator then changed to another set up file and raw positioning was output instead of a position in the user's datum. This simple unchecked error resulting from a change to the system used at the check, led to the placement of about 350 Te of rock some 125 m east of where it had been required. Fortunately, the error was picked up quickly; no rock had been placed on third party assets and there was sufficient rock in the hold to complete the work scope. It is a lesson in making certain that position checks are conducted upon the system to be actually used to control the SRI operation. Personnel must also be alert to any distinctive seabed features for further confirmation that the rock is to be placed at the correct location. The incident described resulted in both wasted time and rock, which is bad enough.

Consequences of set up errors

A worst case outcome would have been the need to return to port for additional rock to complete a small section, or to remobilise personnel and a ship for small remedial works at a later date. The vessel master, contractor and operator's project staff may also be required to make the appropriate reports to the relevant national authorities and the end client's project staff. There is also the prospect of penalties, censure and loss of reputation.

On a more serious note, if the vessel is using an over the side or stern rock placement system the DPOs (dynamic positioning operators) must set the DP system to pivot around that point, especially if working close to platforms or wind turbine towers. Once again, all personnel should be alert to reference points in use and the ship's dimensions (beam and draught) may be altered by deploying an inclined fall pipe.

The bridges of most rock installation vessels are large areas designed to enclose ship navigation, control, survey and SRI operations. The sheltered air conditioned area may extend across the full beam of the ship and bridge staff may not get a true feel for ambient conditions. Going out onto an exposed bridge wing or bridge roof gives a real indication of ambient conditions.

Bridge personnel

A good bridge team instinctively uses all the information available to them. Rule 5 of the International Collision Regulations require all vessels to keep a proper look out by sight and by hearing as well as all other means available.

DPFPV *Simon Stevin* bridge from starboard side (Reproduced with kind permission of Jan De Nul Group)

FFPV *Stornes* working close to an offshore substation (© Van Oord and reproduced with their kind permission)

It is too easy to place an over reliance on the navigation screen provided by the survey team. The term navigation screen when used to refer to a repeat of the survey screen when on location can be seen as a misnomer. It is a repeat of the survey screen with a field drawing and berm references. The survey navigation screen does not display bathymetric and other information.

Other sources of information are the navigation chart, either electronic or paper, radar, VHF radio, CCTV. Alertness to all sensors and a seaman's eye exercised by experienced bridge watchkeepers remains as vital as ever in an increasingly digital ship's bridge. Rock installation vessels are big and heavy. Distances appear to be much smaller when in a large vessel close to assets.

Vessel operating limits

The marine environment is harsh and unforgiving. Those managing the offshore phase of projects have to be familiar with meteorology and oceanography at sea. The project team should have a thorough understanding of the effects of bad weather on a ship, personnel and its implications. All ships have a safe transit envelope and a smaller safe operating envelope in which to work.

An operating limit is often quoted in terms of significant wave height and wind speed. This limit is to provide guidance for both the marine crew and project personnel; to aid planning the management of the ship and work scope in marginal and adverse weather conditions.

Wave front direction and wind direction may be coming from different directions. A confused sea will prevent the DP system from finding a solution to enable the DPOs to keep the ship in position with the precision required. Acoustic positioning systems are affected by ship movement in poor weather. Vessel motion can cause white noise and aeration over the transducers resulting in the loss or degradation of the acoustics. Therefore the ship may not be able to install rock with sufficient accuracy in the target area if its acoustic navigation is degraded, or if the DP system is working hard against unfavourable sea and wind conditions to maintain position.

The DP system measures by means of anemometers the wind force acting upon the vessel. The system then computes the other forces as the DP current which is an aggregate of other forces on the vessel such as tidal stream, current or wash from other nearby vessels. A DP system cannot anticipate swell and vessel movement

Satellite precise navigation systems have brought high standards of positional accuracy to offshore and other marine operations. Users must be aware that poor satellite geometry and extreme low or high satellite elevations will degrade positioning quality. Degraded positioning quality caused by a poor satellite constellation is exceptionally rare now, but more likely to occur in higher latitudes.

Sea conditions

There are no hard and fast rules to gauge when sea conditions will moderate following foul weather. In deeper water, once the wind speed drops below 30 knots, it is reasonable to expect sea conditions to moderate to enable work in between four to six hours. In shallower and exposed nearshore waters residual swells resulting from bad weather can take much longer to subside.

When an optimum position for DP operations has been found vessel movement may prevent certain equipment from being safely operated. With the exception of heave, vessel movements of pitch, roll and yaw are all angular motions. Therefore the angular distance increases from the pivot point, which is normally about one third of the length of the vessel from the bow and below the water line.

Whilst it is very frustrating awaiting sea and wind to moderate to allow work, sometimes conditions mitigate favourably to enable the vessel to continue to work through adverse conditions. The vessel master and superintendent will always be aware of marginal weather and sea conditions and will take the appropriate action, especially faced with a deteriorating weather forecast or conditions.

Rough weather in the North Sea (westerly in excess of 50 knots – storm force 10)

A simple wisdom is that it is more difficult to recover equipment than to deploy it. Self-inflicted damage to equipment caused by an overzealous drive to start work can result in days of downtime and its unwelcome consequences.

Crew fatigue

Crew fatigue can become a hazard during and following a prolonged period of rough weather. Vessel motion, vibration and noise all combine to prevent personnel from having sufficient rest and make moving around a ship difficult. A ship should be sea fastened to prevent anything from breaking loose and becoming a hazard to the ship, project equipment and personnel. The most extreme conditions present a considerable risk to the ship. It is a far safer option to remain in sheltered water and wait for bad weather to pass, than to be hove to into bad weather on site for days, waiting for the weather to calm down. The master will plan to arrive on site to coincide with conditions improving to become workable although it may be necessary to sail through approaching bad weather to achieve that.

Ocean currents

Ocean currents can adversely affect deepwater continental slope operations. Eddies circulating from the North Atlantic Drift are often encountered in mid-water c. 150 m to 250 m, West of Shetland. A current flowing in a north easterly at a rate of two knots or more is not uncommon. A strong current can force the suspension of SRI operations as the fall pipe cannot be kept in position and the whole system is at risk of being placed under some stress. It is possible to lose a number of hours per day to strong mid-water ocean currents when working in deep water West of Shetland.

Environmental considerations

Environmental considerations cover both the wider environment in which the vessel is working and the ambient conditions in which personnel are required to work. Projects are required to submit an environmental impact study to the relevant authorities before any approval for work is given. There are medium to long term benefits which may be considered, such as rock berms becoming an artificial reef.

The safety of personnel is always of paramount importance and must be an integral part of any project. The Level 2 risk assessment is a living document and needs to be formally discussed and revisited as ambient conditions change. For example, loading rock on a sunny calm afternoon is a very different undertaking from loading rock overnight in Northern Norway in midwinter with sub-zero temperatures and high winds. Quarry loading

Katabatic winds producing rough weather alongside at Sandnessjoen (January 2011)

Summer Loading at Bremanger with the vessel on DP, June 2020

systems have operating limits and in severe conditions, the quarry staff and ship's staff stop loading for reasons of safety. Even sheltered waters can be rough when weather conditions and topography combine to produce strong katabatic winds. This happens across a narrow normally sheltered fjord in N Norway where the katabatic winds reached 60 to 80 knots. Fortunately, at that location the katabatic winds blow the vessel onto the wharf.

The contractor and client must conduct their operations responsibly with regard to the wider environment. Closer to home the appropriate PPE must be available for all personnel, but especially those exposed to extremes of weather on deck.

Cold weather

Despite the experiences of working West of Shetland and a winter in Baku, the author was surprised by the extent of cold in northern Norway over a winter. Extra layers of clothing were needed to remain comfortable whilst on deck for any longer than two or three minutes. The contractor found out the hard way that west European winter diesel used to power the excavators became waxy and unusable and it was soon replaced by Arctic diesel.

Frozen quarry, February 2010

The problems of cold weather can impinge upon quarry operations. While dust levels are controlled by dampening the rock with water in summer, in winter rock spraying carries the very real risk of freezing the rock stockpile together. A frozen rock stockpile will delay loading operations until it can be thawed out, which is a matter of some days. Quarries in the north of Norway, where there is a risk of freezing have heaters placed in tunnels at the base of the stockpile to keep the rock ice free. The author's project was delayed by two days because the heaters had failed and the rock pile was frozen solid. The downside of dry loading is that windblown dust can carry a long way and it is a nuisance to neighbouring residents, properties and businesses.

Operating in the tropics

In tropical and subtropical regions, the main risks are from the heat which exposes personnel to dehydration and sunburn and equipment to overheating. Modern vessels are designed with air conditioning which will be working hard. Following an air conditioning failure, a ship will heat up very quickly and it then takes days for temperatures to return to a comfortable level.

The risk of sunburn and dehydration for exposed personnel should be included in the risk assessment and toolbox talks. Sunblock and water should be readily available. Tropical sun is strong and sunburn can affect unprotected fair-skinned people in less than an hour. Even in a North Sea summer, exposed personnel are at risk of sunburn. Daylight hours are long and on clear days the sun can be strong.

Catering staff will be advised to wash fresh food, especially fruit, to reduce the risk of spreading any gastric problems. The author has very sour memories of bad milk that was embarked before a tropical ocean crossing. It had been stored unrefrigerated on the jetty in a subtropical area and therefore became unusable.

Malaria is endemic in some tropical and sub tropical regions. The appropriate antimalarial precautions including preventive medication have to be provided. The use of antimalarial drugs is not without risk either with potential side effects ranging from headaches to more debilitating conditions. The risk of other insect borne and locally prevalent disease should be taken into account during the HIRA.

The marine habitat

HSE protocols require minimal harm to the environment, but strive towards no harm to the environment. Subsea assets, such as pipelines, provide corridors of refuge for marine creatures and support ecosystems. Some will support numbers of top predators, such as ling. Five hundred metre safety zones around floating production and storage offshore (FPSO) platforms and other structures also provide areas where marine life is free from fishing activity. Artificial reef conditions are often created in pipeline corridors and other subsea structures, including rock berms.

Casual and anecdotal observations point to rock installation having an overall medium to long term benefit to the marine environment by providing a habitat. Rock berms act as artificial reefs and provide anchors for sedentary soft bodied creatures like sea anemones and crustacea such as barnacles and bivalve molluscs. This in turn attracts mobile crustacea such as prawns, crabs and lobsters, mobile molluscs, small and juvenile fish.

Smaller predatory species soon exploit a source of food and then larger predators move in. During one deepwater development in Norway, few fish were observed during the pipeline pre-lay, as-laid and as-built visual surveys. A baseline survey of the whole field was conducted within a year of the last rock berms being installed. During the course of the survey seven species of fish were frequently seen, including large top predators such as halibut, ling and tusk.

On several occasions the author has observed large numbers of coalfish or saithe between the MBES (multi beam echo sounder) transducer and the seabed such that both the MBES and visibility were completely degraded. The gas-filled cylindrical swim bladders of these fish form excellent reflectors of MBES waves and their large numbers meant that both the pipeline and seabed were masked from the multibeam. Fine silt from fish foraging on the seabed can readily be stirred into a suspension which can entirely degrade visibility. The standard method of regaining some control is to stop the ROV and turn off the lights for a time. However, within minutes of resuming work, fish will return again in large numbers to take advantage of the light.

During a recent visual inspection of older UK assets in the northern North Sea, the author observed nine species of fish whilst viewing ROV cameras. These included saithe, Pollack (or Pollock), redfish, cod, haddock, angler fish or monkfish and wolffish (*Anarhichas lupus*). The numbers of wolffish have declined dramatically due to a combination of overfishing, bycatch (fish caught inintentionally) and habitat loss. Slow maturing and slow breeding long-lived species are inevitably at the most risk.

There are two schools of thought covering the impact of the offshore industry upon the marine habitat. One school subscribes to the benefits of leaving the artificial reef conditions alone long after the oil and gas field has been decommissioned, while another advocates the return to a pristine pre-development seabed.

Some areas of offshore activity are important migration routes for cetaceans and may require a marine mammal observer (MMO) to be embarked. West of

A wind generator tower from DPFPV *Simon Stevin*

Shetland one can frequently see fin whales, orca, pilot whales and several other cetacean species. Additional measures to avoid disturbing marine mammals with noisy activity and acoustic instruments may be required.

3 Quality control and quality assurance

As is the case with safety considerations, quality control has to be designed into every project from its inception. The end result will be influenced by the quality and quantity of information the contractor has at their disposal for the planning phase. Short duration jobs can be frantic as the contractor may well be completing the reporting stage for the last client, carrying out a job and preparing for the next client.

Quality control factors include but are not restricted to the following points:

- An overall due diligence.
- Choosing the appropriate vessel to meet the requirement. Availability of vessels, commercial and operational constraints may make this difficult and compromises may be necessary.
- The HIRA can be laborious; however it is the opportunity to look seriously at a risk assessment with fresh eyes. The quality of the HIRA may hinge upon the experience and proactive approach of a wide and varied attendance and participation.
- Choosing the correct materials and gauge of rock for the task. This is important and any queries should be discussed with the contractor at an early opportunity to ensure that the rock gauge matches the task.
- The client should provide as much bathymetric and other relevant field data as possible in an agreed digital format to the contractor. This is essential; the contractor will need useable digital data for planning the detail of the operation.
- Providing information about the seabed conditions, if known. Geotechnical data is often sparse; it is not

unknown for considerable volumes of rock to sink in a soft seabed.

- Specifying the rock berms as comprehensively as possible in terms of top width, vertical tolerance, depth of cover and side slope gradient. It is important to ensure that the specification matches the rock size. If in doubt, a dialogue should be held with the contractor.
- Avoiding changing specification criteria whilst the rock installation vessel is on task. It may often prove to be unavoidable for post-lay stabilisation berms.
- Vessel assurance. All client companies need to have evidence that the vessel is in date for maintenance and other classification requirements. The vessel assurance will answer the following questions amongst others;
    - Is the vessel fit for purpose?
    - Does it comply with relevant regulation?
    - Is equipment inspection and certification in date?
    - Are staff levels and the staff skill mix sufficient for the job?
    - Do personnel hold the relevant up-to-date certification?
- Checking that the bore of the fall pipe is compatible with the rock size for the project; for example, a smaller bore fall pipe and a large rock size presents a high risk of fall pipe blockage and further damage.
- Quarry quality assurance – this is especially important in new and smaller quarries and when using a vessel with a smaller bore fall pipe. Larger and oversize rocks will block a fall pipe causing delay and increasing the risk of damage to the vessel equipment and to subsea assets. Smaller quarries may not be permanently working and staff may be temporary. A change of quarry ownership or management structure may also require additional assurance that the rock is being prepared to specification and protected from contamination from oversize rock.
- Keeping the contractor advised of any simultaneous operations.
    1. Some subsea works and rock installations are incompatible and neither party will welcome avoidable delays to the project. Fine rock debris from rock placement can travel up to 2 km and a cloud of fine rock dust can take

some hours to clear. This will degrade visual operations for ROVS and divers and would contaminate any open faces during time in operations. Personnel working on board rock installation vessels should be wary that localised degraded visibility may also be caused by ROVs landing heavily on a soft seabed. A rock installation vessel is an easy scapegoat.

2. The operator should inform the contractor of any seismic vessel activity during planned SRI operations. Seismic vessels tow long streamers and a rock installation vessel with its fall pipe deployed means that both have very restricted manoeuvrability.

- Management chain – Offshore personnel must be fully briefed on the management structure and they have to comply with it. A system of regular conference calls should be in place.
- Any instruction given to the contractor or subcontractor must clearly state that it is an instruction and outline what is to be done. For example, 'the named subcontractor is to carry out the following works or demobilise from the works at a definite stage'.
- Specifying vessel system checks. It is essential that the contractor can demonstrate that all the systems are working within specification in order for rock to be placed where required.
- Specifying deliverables. The standard deliverables are reports, listings and graphics.
- The client should specify the scale required for any graphics. See note below

Note

All interested parties should be fully aware that the horizontal scale for graphics may be 10 times smaller than the vertical scale, so there is an exaggeration of the vertical scale in cross profiles and in longitudinal profiles.

Typical horizontal scales for graphics are 1:1000 and 1:500 where 1 cm represents 10 or 5 metres on the ground respectively. If the same vertical scale value were to be used for a relatively small vertical difference; the graphics would be difficult to resolve by eye. Therefore, a vertical scale of 1:100 is generally used. At a scale of 1:100, 1 cm represents 1 m, which is easy to observe and check that the vertical coverage meets the specification.

In general, the average person can satisfactorily resolve about one third of a millimetre by eye. For example, a 3 metre sinuation in a contour line caused by fluctuations in navigation presented at a large horizontal scale of 1:500 would appear as a wobbly line with an amplitude of 6 mm, at a smaller scale of 1:10000 the line would appear to be straight.

4 Brief summary of the law of the sea

International law of the sea is complex and lawyers and consultants with working knowledge are sought after by nation states and corporate bodies to give advice.

The account presented here is a very brief summary of the law of the sea to make readers aware and to ensure they seek the appropriate advice in the planning stage of a project. International maritime law is based upon the 1982 United Nations Convention on the Law of the Sea (UNCLOS). Most countries have ratified UNCLOS, notable exceptions include the USA, landlocked Uganda, Turkey, Kazakhstan and Turkmenistan, the last two of which have coastlines bordering the Caspian Sea. UNCLOS lays down a coastal state's responsibilities, jurisdiction and its rights to regulate the exploitation of minerals and fisheries. It also defines the responsibilities and jurisdiction of vessel flag states.

Where territorial waters and EEZ (Exclusive Economic Zone) of neighbouring countries would overlap, a median line is defined by mutual agreement or arbitration, if required. Controversial issues include the baseline for any median line and the use of land reclamation. At the time of writing Singapore and neighbouring Malaysia were in negotiation to change the median lines following land reclamation by Singapore. The acquisition of a small area of reclaimed land may confer a huge advantage in extending a nation's EEZ.

The following types of water body are covered by UNCLOS:

Internal waters e.g. rivers, lakes, estuaries and the landward side of enclosed seas such as bays, for example the Thames Estuary, St Brides Bay in Pembrokeshire or groups of islands forming archipelagos e.g. Hebrides, Philippines. Internal waterways are governed by the coastal state's national law and the right of free passage does not exist within them.

Territorial waters up to 12 miles from the baseline which may be from a bay or archipelago baseline or from the mean low water mark. The coastal states law applies within its territorial waters and foreign vessels including war ships have the right of innocent passage. Beyond a country's territorial waters there is a contiguous zone extending up to a further 12 miles for the territorial waters. A state can exercise jurisdiction in the contiguous zone for infringements that are suspected to have occurred within its territorial waters.

Exclusive Economic Zone (EEZ) from 12 up to 200 miles from the territorial waters baseline. Coastal states can control mineral exploitation and fisheries within their EEZ. In the EEZ the coastal state law applies in some circumstances, in others a vessel's flag state law is applicable. Countries can only declare an EEZ if their coastal mapping and nautical charting is of a defined standard. A country may apply to extend its EEZ to the extent of the continental shelf (c.200 metres depth).

International Waters also known as the high seas. No country can claim sovereignty in international waters. Flag state law applies to ships. International waters start from the territorial waters limit of 12 miles.

The International Seabed Authority is an intergovernmental body set up to regulate and control offshore mining and related activity outside EEZs. This is a controversial area as so little is known about the deep sea environment. Within a short period of time, commercially exploited deep water fish species, such as orange roughy (*Hoplostethus atlanticus*) and Patagonian tooth fish (*Dissostichus eleginoides*) with slow maturing and breeding rates have been severely stressed by industrial and often illegal fishing activity.

5 A short history of subsea rock installation (SRI)

Subsea rock installation began in the 1950s in the Netherlands as part of land reclamation and coastal engineering works. Small barges were adapted to place rock in shallow water to provide stable foundations for land reclamation projects. Other uses of rock were to protect pipelines, outfalls and to provide support for structures. Small powered vessels were also developed to carry out this work.

The coastal engineering and inshore market remains viable and a number of side stone dumpers (see page 52 for further information) are in service for coastal engineering and very shallow water inshore projects. Side stone vessels are typically small with a shallow draught. The rock capacity of a typical inshore shallow water side stone dumper is in the region of 1500 to 3000 tonnes (Te).

Unlike coastal engineering in shallow coastal water, the North Sea oil fields were developed in much deeper open water and up to hundreds of nautical miles from a quarry. The vessel characteristics for offshore rock installation vessels were and still are:

- a deep draught and good seakeeping qualities
- a cargo capacity of 10 000 Te or greater
- the ability to keep station and to deliver rock to the required area
- the ability to survey the site and monitor progress to meet the specification

The first offshore rock installation vessels were converted from bulk carriers or heavy lift vessels.

In order to achieve the successful execution of a project far offshore, a number of developments were necessary to improve positioning accuracy and to control a ship precisely. Many fields were too far from land to benefit from higher frequency precision positioning systems that were in common use in coastal waters and instead they relied upon lower frequency coarser positioning systems. To try to improve accuracy, local networks were established within oil fields to bring the benefits of more precise positioning systems offshore.

When working with the lower frequency radio positioning systems, the operator had to know the vessel position, reasonably accurately, in order to set the receiver. This could be done by passing close to a known point, finding a subsea feature or using another more accurate radio positioning system to seed it when closer to shore. The lower frequency long range systems were very susceptible to atmospheric changes and interference, especially at night and around sunset and sunrise. It was not unusual for positioning to degrade severely and sometimes go unnoticed. Position line geometry was also most important and position lines which crossed at less than 30 degrees or more than 150 degrees were to be avoided whenever practical.

In the late 1980s, the Global Positioning System (GPS) satellite navigation system entered service. It was owned by United States Department of Defense (US DOD) and developed primarily for the use of the US armed forces. It operated using two frequencies to give good accuracy. Only one of these frequencies was made available to civilian and commercial users and random fluctuating errors were set up for non US military users. Companies were quick to see the benefits of this new system and set up highly accurately positioned differential stations onshore to monitor the errors, then transmit the corrections to users via a satellite link. In general users had to be within about 500 km of a differential station and be receiving data from the same satellites. In its infancy DGPS (differential GPS) signals were sent via HF radio links, but this was a very short-lived method of delivery. GPS was a huge change to all interested parties and accuracy previously only available within 40 to 50 km of the coast was possible several hundred kilometres from a shore station. With a few exceptions, radio positioning aids were nearly all superseded when differential GPS coverage became widely available with close to 24 hour coverage by late 1992. The introduction of GPS changed everything. It now brings high standards of accuracy to most parts of the world.

Positioning was just one of a number of technical, engineering and logistical challenges that had to be overcome. As positioning became much more reliable and offshore rock installation was viable, other challenges could be addressed, such as:

- quarry facilities and infrastructure for efficient loading and turnaround of vessels
- the capability to deliver rock accurately and safely to the target site at depth

Simon Stevin, the world's joint largest rock installation vessel at Skarv Field, April 2011, showing the inclined fall pipe in the stowed position (Reproduced with kind permission of Jan De Nul Group)

- the need for a significantly larger cargo capacity than that offered by the side stone dumpers
- the need to develop techniques to avoid damaging the product and infrastructure at all costs

In recent years the need to keep within permitted and licensed limits has to be most carefully monitored and with modern positioning systems that can be readily managed.

The requirement to be able to place rocks at greater depths has led to the development of the fall pipe as the integral part of a rock installation system. This is examined in more detail in later chapters. The fall pipe channels rock to where it is needed avoiding a widespread distribution.

By 2010 the main rock installation contractors had new purpose-designed vessels on order or under construction. Since 2011, eight new rock installation vessels have entered the market, whilst only two older vessels have been retired from service. These complex and highly technical vessels represent a significant investment by and commitment from the contractors.

Five of the eight newer vessels have a maximum capacity in excess of 23 000 Te, and of these two have a capacity exceeding 31 000 Te (see table in Appendix B). The four current main rock installation contractors are Boskalis, DEME/ Tideway, Jan De Nul Group and Van Oord. These are all large companies based in Belgium and the Netherlands, where there is a long history of expertise and innovation in all aspects of coastal engineering. Subsea rock installation is just one area of expertise for these large important companies. Their other activities include dredging, land reclamation, civil engineering, offshore oil and gas field development, telecom and interconnector power cable support and more recently offshore renewable energy development. With the need to develop specialist vessels, they also support marine industries and can drive innovation, particularly in the use of cleaner fuel.

6 Purpose of and uses for SRI

Subsea rock installation is self descriptive and it is part of the general engineering offshore for:

- oil and gas infrastructure
- telecom cables
- power cables
- offshore renewable infrastructure.
- civil engineering

Subsea rock installation is carried out primarily to protect, stabilise and support subsea structures, pipelines, telecom cables and power cables. To eliminate the risk of assets sinking into a soft seabed, a rock berm can be placed in those areas to spread the weight of an asset, provide support or to facilitate the wet storage of components and equipment placed on the seabed for later use and eventual recovery. Rock berms can also be used to cover unused or abandoned well heads and other surplus infrastructure. Very occasionally a rock berm can be used to contain an area of interest for subsequent action.

A rock berm is a particularly useful structure to provide a stable and relatively clean storage area for equipment and tools to be used during tie ins and other similar operations being carried out on a soft or fine seabed.

Rock installation can also be used to provide support pads for jack up drilling rigs to be located in areas of unstable seabed soil. The rock pads spread the load and reduce the risk of the seabed sheering under the weight of the jackup rig. Jackup rig operators require very precise level tolerances on any support pads. This can be partly offset by the fact that jackup rig legs will displace some rock

Stabilisation berms are generally put in place in addition to a protection berm to prevent catastrophic upheaval and lateral buckling of oil, gas and water injection lines that are operated at a range of variable temperatures and pressures.

Legal framework

Offshore activity takes place in a framework of legislation and its associated regulations. The rules and conventions are after international and national, examples include; UNCLOS, the International Collision Regulations, The International Shipping and Port Facility Regulations (ISPS) and the International Labour Organisation (ILO) Marine Labour Convention of 2006.

States will control shipping regulation and the certification of seafarers. National regulatory authorities have the right to inspect vessels for compliance.

Offshore activity in the UK Exclusive Economic Zone (EEZ)

Regulatory authorities may require subsea oil and gas pipe lines, umbilical lines and cables located on the continental shelf area to be buried to a certain depth to mitigate against the risk of damage from trawls, impact from dropped objects, anchors and anchor cables. In the UK the regulations come from several sources. The relevant UK legislation and regulation includes but is not restricted to:

The Pipelines Act 1962

The Petroleum Act 1998

The Energy Act 2008

The Health and Safety at Works Act 1974

The Electricity and Pipelines Works (Assessment of Environmental Effects) regulations 1990

Pipelines Safety Regulations 1996 and the accompanying book *A guide to the Pipelines Safety Regulations*, 1996 published by UK HSE

The Submarine Pipelines (Inspectors and Safety) (Amendment) Regulations 1991

The Offshore Petroleum Production and Pipelines (assessment of Environmental Effects) Regulations 1999 as amended

Depending upon pipeline diameter and operating parameters, all pipelines less than 16 inches in diameter are required to be buried unless they meet exemption criteria. Cables should be buried to a depth of one metre to afford protection from damage by trawls.

Offshore activity in the Norwegian Exclusive Economic Zone (EEZ)

Norway has the world's fourteenth highest oil production and is seventh in exports. It is Europe's largest producer and exporter. The country has a highly developed regulatory structure and its industry and regulatory bodies are world leaders in many aspects of technology and management.

The legislation is enacted in the Petroleum Act of 29 November 1996 and the following regulatory bodies are established to oversee offshore oil and gas industry activity:

The Ministry of Petroleum and Energy (MPE)

The Ministry of Labour

The Ministry of Finance

Two further regulatory bodies have a day-to-day management role:

National Petroleum Directorate (NPD)

Petroleum Safety Authority (PSA)

Quarrying in Norway is covered by the Norwegian Mineral Act.

Bribery legislation

Many nations have anti-corruption legislation covering the standards and ethics of a company's conduct with regard to its operations worldwide. UK companies have to demonstrate compliance with the Bribery Act 2010 (amendments in 2017). The rules cover the provision and receipt of hospitality, gifts and cash payments in a range of circumstances from tendering to facilitation. A $100 cash payment to a local "Mr Fix it" is corruption, a $10,000 add-on disguised in the invoice is commerce.

Rock installation or seabed burial

Project management teams often have a choice between ploughing, trenching and rock installation to provide protective cover and stabilisation. Legislation and regulation may favour a particular method, and as always cost and availability will be a major consideration. In general rock installation is a premium product, which will deliver the required coverage to the client. It should be borne in mind that both trenching and ploughing may need to be supplemented with remedial rock installation, especially if the cover depth is not achieved and stabilisation or upheaval buckling mitigation is required.

By its very nature remedial SRI can be time consuming and will lead to a higher than expected volume of rock use. The vessel staff have to balance accuracy against vessel time and the risk of having to embark additional loads. The offshore representative and client company will naturally steer the vessel staff towards accuracy, especially if line testing is dependent upon the completion of SRI.

Trenching and ploughing

Trenchers and ploughs are subject to considerable mechanical stresses, vibration and the operating window for launch and recovery has to be carefully monitored. The efficacy of trenching and ploughing is also seriously affected by unknown or unobserved changes in seabed soil conditions. Changes in seabed soil may require a trencher to be recovered and its buoyancy adjusted to either increase traction or reduce its risk of sinking. Trenchers and ploughs are recovered to deck at crossings. They are then redeployed at a safe distance from the crossing to resume operations. The crossing can be protected by either mattresses or a rock berm.

Recovery of trencher to support vessel

Unknown and unexpected seabed and subsoil conditions can cause delay. In the worst case it is a driver for a significant post-trenching phase of remedial subsea rock installation works, especially for upheaval or lateral buckling mitigation. A soft seabed can make trenching and ploughing difficult as additional buoyancy may be needed on the trencher, which can cause traction problems as a buoyant trencher skips over the seabed. On the other hand, a hard stiff seabed such as clay, will reduce the trenching speed, sometimes to as low as tens of metres per hour. Trenching and ploughing requires considerable management and the right application of long-acquired practical, technical and experiential skills.

Fishing activity

In general, larger diameter pipelines can present a hazard to fishing vessels and their towed gear, whereas fishing activity is a hazard to smaller diameter pipelines and cables. Ideally pipelines should be well charted and fishing vessels should keep clear of them. It is a fact that pipelines attract fish, often in commercially exploitable quantities. A trawl can drag cables for tens of metres and may damage or completely break a fibre optic cable if its critical bend radius is exceeded.

Fishing communities are under immense pressure, with high vessel overheads, catch quotas and restrictions upon days at sea. Anecdotal reports suggest that the crews of some fishing vessels have adapted their nets to enable them to fish very close to pipelines. During his offshore time the author has very rarely seen substantial amounts of fishing gear on pipelines and structures, or higher levels of fishing activity close to pipelines. The idea of adapted nets is either an urban legend or a very successful boost for a select few amongst many hard pressed fishing vessel operators.

Shallow water renewable developments have a large and immediate effect upon fishing communities as the developments become a no-fishing area. The benefits of rock berms and towers acting as artificial reefs and fish breeding grounds takes some time to be realised.

Technological improvements

Further developments include the continued improvement of GPS and similar systems for surface positioning and control of dynamic positioning systems. Subsea acoustic positioning and multi beam echo sounders, with software have together contributed to extending the possible developments and uses of subsea rock installation.

Engineering and innovation

Rock installation applications have grown to include preparation for field development, reducing gradients and preparing crossings, pre-construction supports, scour protection, filling in excavation areas where natural backfill is unlikely, covering disused structures and coastal civil engineering functions such as breakwater construction.

In shallow water and in areas of stronger tidal streams rock is used to provide scour protection for wind generator towers, offshore substations and other infrastructure. Denser rock, such as eclogite, is chosen for works in these areas.

New uses for rock installation are being developed, for example a rock berm can be used to assist in a number of coastal engineering applications such as land reclamation, building breakwaters and artificial reefs. Rock installation may be both the first and the final phases of an offshore field development project.

Innovators will develop new methods and adapt tried and tested methods to solve new problems and to optimise vessel time. Quarry turnaround times can be reduced depending upon the design and compatibility of the loading system with a vessel's layout and hull design.

7 Industry requirements of SRI

In the oil and gas sectors rock placement is most closely associated with pipeline and umbilical protection and pipeline stabilisation. In the late 1970s the need to protect and stabilise pipelines servicing North Sea oil fields emerged. Subsea oil and gas pipelines operate at temperatures above the ambient levels and steel expands when exposed to higher temperatures. Crude oil and gas recovered from a well is hot and may contain up to 20% sand and water. The sand and water are removed before the cleaned oil is transported to shore via a pipeline or shuttle tanker. The residual or produced water can be pumped back into the well to maintain pressure and production levels. Produced water is hot and the line will be pressurised, therefore water injection lines may need stabilisation to mitigate the risks arising from the effects of both increased temperature and pressure. As the well gets older the temperature of the produced water tends to increase and this is a factor in designing water injection stabilisation.

The operating parameters and physical characteristics of an individual pipeline, which include composition, wall thickness and resting position, are used to calculate the likelihood and locations of potential upheaval or lateral buckling. Where the risk is identified it can be mitigated by placing a weight of material on top or to the side of the potential buckling points. Care is also taken to check that the risk of any buckling is not transferred to other locations. In some cases, stabilisation may be achieved by trenching and natural backfill, however in a significant number of cases subsea rock installation is used to complete any lateral or upheaval buckling mitigation. Buckling is often angular with both lateral and upheaval elements.

In the UK exclusive economic zone (EEZ) burial in the seabed is the preferred method of pipeline protection. Contractors and operators will have to make a case for rock installation as an alternative to burial. Subsea soil conditions to be found in the eclectic mix of glacial deposits that make up much of the seabed in the North Sea, can be difficult for both plough and trencher operations. Therefore rock installation can provide a viable alternative to afford protection.

Initial planning

The scope for SRI should be included in the initial project planning. The earliest phases of planning should always include a review of charting in the area of interest. Choices in route selection may be restricted by other structures, the nature of the seabed and seabed topography. For example, it is worthwhile to check gradients carefully and if the product is to be placed parallel to the contours of a slope, assess the risk of it slipping down the slope. An alternative route may be necessary or an additional rock berm could provide a shallower gradient or supporting shelf. Unsuitable seabed topography and seabed soil conditions can present a very real risk to the project, if not thoroughly understood and mitigated in the early stages.

Cold water coral and other benthic colonies, which are found in deeper water on the ocean margins, are protected and must be avoided. The design pipe route has to be specified to keep clear of them, where known and if cold water coral or other benthic colonies are found during route and site surveys, further modification to the route must be made.

Purpose of the berms

The project team should fully identify the purpose of rock berms and what SRI is needed at various stages of the project. It should be noted that the additional quantities of rock required for stabilisation above the protection requirements will be derived by pipeline specialists using the post-lay out-of-straightness (OOS) survey data. The analysis will identify points in the line that would buckle if placed under heat and or pressure stress. Rock installation to stabilise the line must be completed before full pipeline pressurisation checks are carried out. The calculations and analysis required to work out the quantity of additional rock for stabilisation are long and complex.

The just-in-time approach can lead to a gap where it may be of overall benefit to the project to suspend the offshore SRI phase. It can then resume following the OOS survey analysis and the issue of a finally accepted plan for the pipeline stabilisation. There is a danger of accepting an initial figure and then using too much rock where it is not necessarily required. The other risk is placing insufficient additional rock and then subsequent remedial works become time consuming rectifications.

The operator may choose a third party to evaluate the OOS data. Such action may lead to contractual issues regarding the liability for any future defects arising from upheaval issues.

Tolerance

In most client specifications, tolerance is expressed as a value with an addition e.g. 1.0 m (+0.0 to 0.4 m). It provides a minimum value required and an acceptable level of excess. There is a relationship between the size of rock, the tolerance that can reasonably be expected and the minimum depth of cover that can be achieved. The larger the rock size the higher the minimum rock berm should be and as a result the tolerance has to be larger.

For berms requiring a level tolerance on the top plane, selection of the correct rock size can be the difference between success and failure. Heavy structures can displace some rock, forming a berm, so a level tolerance may be met by sheer accident of rock displacement.

In general, large size rock will increase the tolerance value that can be achieved and the minimum quantity of rock coverage that can be placed. A 30 cm high rock berm is the minimum that can be realistically achieved with 1 inch to 5 inch (c.25 mm to c.125 mm) rock. The almost universal standard tolerance for berms built with 1 inch to 5 inch (c.25 mm to c.125 mm) rock is (0.0 to +0.4 m).

Very rarely, a client specification can be unrealistic in requiring a tight tolerance e.g. (0.0 to +0.1 m) with a larger rock grade such as 2 inch to 8 inch (c.50 mm to c.225 mm). A tolerance of 0 to +0.25 m is extremely difficult to achieve with 1 inch to 5 inch (c.22 mm to c.125 mm) rock. This invariably results in frustration and inertia.

This is very relevant in projects where there are very particular level requirements for the product or structure to be placed upon the berm. There may also be engineering limitations upon the weight of rock that can be placed on top of a product.

Pre-lay berm

Some pipelines need to be laid on a near level seabed. A pre-lay berm may be required to reduce gradients over the pipeline route. Where a pipeline is laid on a pre-lay rock berm the standard pipe line lay corridor width may be reduced to ensure that the pipeline is laid in the centre of the berm. If the pipe is laid too close to the edge it may be at risk of falling from the rock berm onto the seabed. This will negate the benefit of the pre-lay berm and seabed topography may compound any problems.

Detailed planning will require hydrographic and geotechnical data and a clear purpose for the subsea rock installation. The project team should also consider the time of year for operations and be mindful of the prevailing weather conditions and vessel operating envelopes.

In deep water, the effect of ocean currents on the fall pipe should be examined. This must be considered when operating West of Shetland when on the continental slope, i.e. deeper than around 200 m. In shallow water the reduced efficacy of acoustic instrumentation and the effects of tidal stream and sea conditions are factors in planning.

The crossing of third party pipelines and cables may be required. From the early stages of a project, the project team will need to identify these lines and start a dialogue to obtain permission from other operators to cross their assets. It is necessary to comply with any third party requirements which may encompass notification, permits, additional reporting and even third party representation offshore. It is wise to know what the third party requires well in advance and to keep the dialogue in progress to reduce any risk of delay arising from the need to negotiate with a third party.

Quantity of rock required

When discussing volumes and tonnages it is of vital importance that all parties are using the same terms regarding theoretical and working tonnages and that everyone fully understands the terms being used. There are important differences which are explained in the following paragraphs.

The quantity of rock required is derived from the calculated volume. Planners will use a theoretical tonnage which is the theoretical volume multiplied by the dry bulk tonnage which is typically 1.55 Te per cubic metre. The actual tonnage required will necessitate a larger multiplication factor to accommodate losses through sinkage, rock settlement, seabed topography and to compensate for the disproportionately higher volume of air in the sample measurement.

First of all, the berm has to be specified in terms of length, width, and amount of cover required over a specified point of interest. If the pipe or cable lies on the seabed the volume required to bring the rock berm level with the top has to be calculated. The simple act of multiplying the volume of rock cover required by the bulk density is a starting point for the tonnage only. Theoretical rock berm volumes may be calculated with the assumption that it is to be placed upon a flat seabed. Seabed topography and soil conditions will have a direct influence on rock quantity by requiring either an increase or reduction in the volume of rock needed to build the berm.

Each company will have a formula for a given project subject to knowing the nature of the seabed and based on collective experience. The actual quantity required may vary from the estimate due to various factors such as, taking into account rocks sinking into or displacing seabed material, a small volume of wastage outside of the specified area or settlement as the rock berm forms. Estimates of rock quantity may also be influenced by trying to anticipate any requirement for stabilisation following an out-of-straightness analysis.

To obtain a realistic estimate of tonnage required for a given volume, a multiplication factor ranging from 2.2 to 3.0 Te per cubic metre of the theoretical volume provides a reasonable figure based upon readily achieving coverage within the tolerance. Smaller berms are usually estimated using a slightly higher multiplication factor. With a one metre crest width and a berm height of one metre and a tolerance +0.0–0.4m there is a 35% difference in volume between the upper and lower tolerance levels. It is easy to forget that an increase in berm height has a corresponding increase in berm width.

Duration

The contractor will schedule the job based on previous experience for similar works with a factor to accommodate for any loss due to weather. Other features such as typical quarry turnaround times, distance to and from the designated quarry, any fall pipe adjustments, in-field transits, and SIMOPS are also taken into account. In the author's experience timings can sometimes be over optimistic. A four to six day turnaround time per load is a good baseline from which to start, depending upon the distance of the work site from the quarry.

UK deposit consent (DEPCON)

In the UK, an oil and gas operator will apply in good time to the Oil and Gas Authority (OGA) for a deposit consent (DEPCON). A DEPCON is required to cover any thing which is to be placed on the seabed and includes concrete mattresses and other items, rock renewables and civil engineering project teams will make an application to the Marine Management Organisation (MMO). The DEPCON has strict geographic and tonnage limits; it is further controlled by a time expiry. Specific parts of a pipeline such as spools, pipeline end termination (PLET) covers and wellhead and redundant infrastructure burials may have individual limits within the DEPCON total. In drafting a DEPCON request, the project team should make a reasonable, well informed estimate to avoid exceeding the DEPCON limit during the project and to avoid an over inflated quantity. The contractor, when selected, should be able to provide an informed estimate of rock usage. It is also prudent to extend the geographic limits of any permits such as a DEPCON to allow for the transition zone to full berm height and differences between vessels' navigation and geodetic systems.

The project manager should avoid requesting over-constraining DEPCON limits, whilst keeping them reasonable. The project team should have a degree of flexibility to cover potential unknowns arising from factors such as seabed conditions and the need for any additional stabilisation berms following post-lay out-of-straightness analysis. The author is of the opinion that a DEPCON should cover an entire field, or lines where multiple lines are under construction. Where practical, a project team should avoid dividing lines into constituent parts such as spools, crossings, etc. especially as the smaller constituents provide the highest risk of over installation. It is possible to be within a total DEPCON limit, but in excess on spools or crossings. The over installation could lead to delays to accommodate amendments or have to be justified post-installation.

A DEPCON is a licence, but it should also be considered to be an operating envelope, which enables the rock installation contractor to complete the works without constraint. However the quantities of rock used should be monitored against the DEPCON. On completion of the works, the operator is required to report the amount of rock used and confirm the geographical limits of the installation to the OGA or MMO as appropriate.

It is far better to overestimate the quantity applied for by a generous amount thus avoiding having to place supplementary requests to either OGA or MMO, or to account for quantities in excess of the limit when the final figures are submitted. A good rule of thumb is to apply for a DEPCON that is at about fifty to seventy percent greater than the calculated tonnage.

In recent tasks the author has noted the following figures regarding DEPCON management

Description	% DEPCON used	Notes
Small berm < 2000 Te	83%	Berm design was changed without a review of DEPCON
Pipeline protection and stabilisation 1	60%	An extension was granted. The total tonnage used was 102% of original value
Pipeline stabilisation and protection 2	78%	Very tight had additional stabilisation been required
Pipeline stabilisation and protection 3	60%	No extension required or even contemplated with plenty of margin for stabilisation had it been required
Two location remedial works	41% 62%	Location 2 was used to empty the rock installation vessel

There is a DEPCON calculation guide in Appendix E.

SRI phase duration

The duration of a project will depend upon:

- the load capacity of the vessel selected
- meteorological and oceanographic conditions
- simultaneous operations

- the quantity of rock required and nature of the seabed
- the turnaround time needed to transit to and from a quarry for each load
- the berm designs and specification
- depth of water and length of fall pipe
- in-field transit distances and any fall pipe management
- additional works arising, such as stabilisation calculated from OOS surveys

A proportion of this time is taken up with the mandatory dynamic positioning (DP) checks, launch of the fall pipe remotely-operated vehicle (FPROV), Survey system checks and calibrations, building the fall pipe, transits, intermediate and post-installation surveys and eventually the recovery of all equipment.

In an ideal world the amount of rock required should be a multiple of nominal whole loads of the vessel selected, although this is rarely the case.

The contractor will build in a weather allowance and it follows that summer is less likely to bring bad weather than winter. However project staff should not be surprised by low pressure frontal systems in summer bringing strong winds and heavy seas, especially from mid-August onwards through until early November.

A table showing typical mission profiles with the percentage of time spent in each activity is provided in Appendix I.

Deposit rates or dump rates

Rock installation vessel deposit rates are very impressive, with newer vessels having the capability of installing up to 2000 Te per hour. The dump rate can be controlled by adjusting the speed of the conveyor belt and the rate of rock release from the hopper. In practice high dump rates cannot be sustained because there are other elements that make up the whole rock installation picture. Taking as an example an average pipeline protection or stabilisation berm, a net deposit rate of 550 to 750 Te per hour is possible from arrival on site to departure. Greater rates can be achieved with a bulk deposit or a straightforward first pass. The rate will be reduced if it is necessary to adjust the fall pipe, make a number of in-field transits and to carry out a number of intermediate surveys.

More complex, smaller berms and the final completion will result in a much lower rate as more time will be spent in relocating and surveying. Additional pre-SRI surveys may be planned when the vessel is empty to enable designs to be processed for use as soon as the vessel returns to site.

The dump rate is also affected by the depth of water and thus the length of fall pipe. Each fall pipe has its safe operating load limit. In practice this starts to feature at depths of around 500 m, depending upon the individual system. At maximum depth the dump rate may be a small proportion of the maximum rate. The safe working load of the fall pipe wires is set by a combination of the winch capacity, the weight of the fall pipe and the friction-generated drag between the fall pipe and the rock water mix within it. The safe working load must also take into account the weight of rock within the fall pipe. Should a blockage occur, rock will quickly back up the pipe and the wires must be capable of holding it to avoid a catastrophic failure. All systems will have an overload alarm and operators may see the effects of a blockage in the upper sections before the alarm indicates a problem.

Therefore, the quantity of rock the fall pipe can hold reduces with its increasing length above a design threshold.

Documentation

A full set of documentation to cover the work in detail is required. It will include:

- A bridging document
- HIRA
- Maps and charts
- Electronic field background information for displays
- Procedures
- A method statement
- Operator company marine procedures

Any drawings and electronic data can be considered to be documentation and should be accounted for accordingly.

In a significant proportion of rock installation cases, the contractor will have a generic procedure, which provides a comprehensive cover of subsea rock installation from their vessel. Rock installation vessels are essentially carrying out a single role and the variables are geodetic details, location, rock materials and the extent and purpose of the rock berm. To overcome the need to repeat the rewriting of a relatively firmly fixed procedure, the contractor's generic procedures can be supplemented with a comprehensive method statement which refers to the generic procedures. The method statement will cover the individual job specifics in detail.

The advantage of embracing this is that the method statement can be published quickly. There is often a short time period between the completion of the as-laid or as-built surveys and the rock placement. There is a sense of urgency to get the pipeline ready for production especially if it is replacing an unusable line.

8 The contract

The contract determines who does what, when and where any demarcation lies. The client company may include SRI within an EPIC (engineering, procurement, installation and commissioning) contract although it is not unknown for a client company to opt to contract rock installation directly.

Fixed price contracts are commonplace, although higher precision and therefore more time consuming rock placement works may be carried out using a system of day rate operations (DARO). The option of using a DARO should be part of the contract for specific tasks such as providing a tight tolerance for height and level of a particular berm or for placing rock to anchor, a PLET cover or similar structure.

Engineering procurement installation and commissioning (EPIC) contract

A client awarding an EPIC contract expects that the contractor will carry out the engineering and pre-planning for the project. The contractor will also start procuring long lead items and subcontracting works. When a timetable has been agreed, approvals granted and equipment and vessels are procured, the installation phase will begin. Offshore construction activity can take place over several years, especially if works are programmed for the summer months only. The contractor must be aware that items designed for subsea use may not fare well if kept for a prolonged period under a tarpaulin in a damp open yard. Finally, the whole project will be tested and commissioned for use before being handed over to the operators.

Contracts are complex legal documents, often open to some interpretation. Many contracts will have a specific confidentiality clause and those that do not, imply it. There is inevitably a gap between what a client expects and that which a contractor intends to do. The team offshore will have to be prepared to deal with reducing that gap in consultation with all concerned.

In the subsea rock installation phases of a project, these gaps can be at their widest when rock installation quickly follows a construction phase. Construction can deviate from the design specification for a whole range of reasons. These include covering overage loops in cables and flexibles and remedial works to bring the depth of cover into specification following ploughing or trenching. Another frequent short notice requirement is for rock installation for upheaval buckling mitigation. The extra cover is calculated from data analysis of the out-of-straightness survey following pipelay. Out-of-straightness data may take seven to ten days to analyse and then to calculate the additional quantities of rock needed.

The contractor companies are operating in a highly competitive market. It is however the case that despite the competition, joint ventures are established to spread the risk and to optimise vessel use and contractor flexibility in big or long duration projects. For example, during the SRI phase of one recent major West of Shetland project, five of the seven available rock installation vessels were working on it for a short period in June 2013.

Other variations in the type of contract are for work to be carried out at a day rate or a net tonnage rate, which would usually exclude any rock placed outside a specified tolerance and the volume of any pipe being covered by the berm, although it may not include any rock that sinks into the sea bed. The net tonnage rate can be further modified according to the type of berm and may even have provision for DARO in specified circumstances.

Projects with a partial load requirement may actively seek the opportunity to combine rock installation voyages with other internal projects or third parties. There is the possibility of an overall benefit of making more efficient use of vessel time and sharing fixed costs such as mobilisation costs.

It is also worth checking the definition of the term 'in specification', especially when reviewing data using a long profile and cross profiles. Some contracts define the specification as the average value of a rolling 10 square metres of rock berm. The values will be derived from the digital terrain model (DTM) gridded to 0.2 m, which gives 2500 spot depth values in 10 square metres. The long and short profiles at 5m intervals presented by the contractor are a representative sample of the whole data set. In critical locations, the DTM should be used to assist in making the decision regarding additional rock installation passes in suspected marginal areas.

The paragraph above is further supported by a recent collaborative paper published by ENSIETA, France, ENSG France and Boskalis Survey Department, Netherlands (Multibeam echosounder error characterisation on dumped rock areas, by N Debese, L Heydal, T Neuman and N Saube). ENSIETA is now ***École*** Nationale Supérieure de Techniques Avancées de Bretagne - National Institute of Advanced Technologies of Brittany often referred as ENSTA Bretagne This paper has initially concluded that MBES typically returns deeper values and the error increases in extent with smaller size rock. It also states that more work is necessary. The results were based on using drying values in a coastal region with a large tidal range as the control. In practice, knowledge of the contents of this paper will have little influence on the average client representative offshore. The client representative is the person who will sign the work off and expects to see a representation of the rock berm to show that the specification has been met.

9 Rock and rock types

Any planning requirement has to involve an understanding of the materials to be used, in this case rock. A basic knowledge of geology, chemistry and electrolytic corrosion is useful. Iron rich rocks may present an electrolytic corrosion risk to steel pipelines and other infrastructure. High pressure conditions in deeper water as shallow as 250 m can cause hyperbaric corrosion of certain steel types.

Rock is defined as an aggregate of minerals and is classified into in three categories; igneous, sedimentary and metamorphic.

The seabed soil has been derived from rock over time and it is what some rock may become over an equally long interval. The nature of seabed soil can affect a rock installation, just as it can any other intervention on the seabed. It is not unusual to find changes in seabed soil along the length of a pipeline and over an area of operation especially in the North Sea basin which has been subject to glacial retreat. Seabed soil is addressed in outline following the discussion about rock. The following paragraphs describe in simple brief terms the different categories of rocks.

Igneous rock

Igneous rock is typically formed from molten rock (magma or lava) that has cooled and solidified after being erupted/extruded at the surface from a volcano (as lava) or intruded into rock surrounding the magma chamber and cooling down to a solid state. Igneous rocks are grouped by their silicon content and also crystal size, with granite and rhyolite being classed as felsic (derived from feldspar and silica), they tend to lighter colours and relatively low density. Mafic (rich in magnesium and iron) igneous rock types include fine grained basalt and coarse grained gabbro, that tend to be darker in colour and denser. Ultramafic igneous rock is denser with a lower silicon content, usually more than 90% mafic minerals.

Sedimentary rock

Most sedimentary rock is secondary in origin, formed from the mechanical breakdown of pre-existing rocks by the agencies of weathering, wind, ice, rivers, stronger tides and ocean currents. Other rocks are formed by chemical or biochemical deposition and some are organic in original. In certain depositional conditions plant and animal deposits may form coal, oil and limestone.

Sedimentary rock types include conglomerates, sandstones, ironstones, mudstones, limestone, coal, chalk and clay. They cover the full range of geological time and were originally laid down flat. It follows that in normal circumstances older rock is overlain by newer rock in a bedded or stratified sequence. However, tectonic activity can tilt them and fold them to form synclines (depression), or anticline (upthrusts). It may also create fault lines in rock, thrust faults can force older rock over newer sediments. Sedimentary rock tends to be of lower density and softer than igneous and metamorphic rocks, although some older sandstones and grits can be very hard. Older sedimentary rocks may be iron rich which makes them unsuitable for rock installation. Sedimentary rock is not currently used for rock installation in the North Sea.

Metamorphic rock

Metamorphic rock is formed from pre-existing rocks due to the effects of high temperature and pressure upon the minerals forming igneous, sedimentary and other metamorphic rocks. These conditions change the original rock mineralogically, texturally and structurally without fully melting it. This can occur at tectonic plate boundary zones (and is called regional metamorphism) and close to magma chambers (thermal metamorphism). Examples of metamorphic rock include slate that is derived from mudstone or shale, quartzite derived from sandstone, marble derived from limestone. Gneiss, schist and some granites are formed by progressively higher temperatures and pressures. The greater changes occur in rock subjected to the highest pressures and temperatures over the longest length of time.

Rock for SRI

Rock for SRI operations is specified by both type and size; either imperial or metric sizes can be specified and each type and size of rock has a use. In general, the client company will specify freshly-crushed granite or schist, both of which are stable and inert. Iron-rich rocks such as basalt, dolerite and ironstone are avoided to prevent any chemical or electrolytic reaction with steel pipes.

Smooth river cobbles

Freshly-crushed 1"–5" rock in a vessel hold

Freshly-crushed rock with a rough surface will lock against adjacent rocks to form stable side slopes whereas weathered and eroded rock, especially smooth river cobbles, will not. Smooth rocks will slide over each other and stable side slopes at the optimum gradients will not be formed.

Finer grade rock and gravel, 1 inch to 3 inch or smaller, can be used as a cushion to protect a product from larger rocks which may form the main rock berm. A larger gauge rock berm placed on top of finer grade rock is sometimes referred to as armour rock.

Many rock types, whether igneous, metamorphic or sedimentary are formed of complex silicate minerals. Denser rock such as eclogite, which is a metamorphic rock formed from mafic igneous rocks deep in the crust in subduction zones, can be used for specific requirements in areas of mobile seabed and strong tidal streams. Eclogite is found at the surface in a small number of locations including Visnes quarry near Kristiansund.

Seabed soil

As much as it is necessary to understand rock and how it can be used in subsea projects, it is important to have an appreciation of the nature of the seabed and what effect it may have upon any structures placed upon it and into it. In the ideal world the seabed will support a structure in the chosen position. In practice some intervention maybe required by the use of a rock berm to spread the structure's weight. Alternatively, structures can be mounted upon piles which extend through the subsoil onto underlying rock.

The seabed is not homogeneous. In some parts of the world a clue to the nature of the seabed can be seen by looking at the adjacent land and topography, in others the links are not so easy to determine. In the initial planning stages the project team should engage with the company's geotechnical department or consultant to obtain as much seabed soil information as possible from previous projects. This may include seabed samples such as cores, grab samples and cone penetration tests, piling and any ploughing or trenching records from the area. Once a general impression is gained additional samples may be necessary to reduce the risks to the project or find the most effective methods. Seabed sampling should be carried out with the route and site survey programme.

Sea beds may comprise sand, mud, clay, boulder clay or rock and several types may be observed in a locality. The project staff may need to know the depth of material laying over bedrock. There are variations in the seabed material; fine material may be compacted and formed into hard stiff clay, or it may be soft and easily displaced. Heavy objects placed upon it are at risk of disappearing deep into the mud. Boulder clay contains rock in a variety of sizes in a matrix of fine material. It can be problematic for trenchers as large boulders can block discharge outlets.

Sand varies from coarse to fine and like mud and clay it can form various surfaces. Sand and mud with a high water content will not support heavier objects. A further consideration with a rock installation over a buried pipe in soft soil is that backfilled soil can form a cushion which will allow the pipe to move. In effect a soft backfill material reduces or even nullifies a rock installation designed for upheaval buckling mitigation as it forms a fluid barrier between rock and pipe.

A rocky seabed may be highly resistant or it can be brittle like the cap rocks found in some areas such as the Persian Gulf. Rock berms placed on a soft seabed may be built using a greater quantity of rock than expected, as a proportion of the rock will sink into the soil. The nature of the seabed and how structures placed upon it will fare is of vital importance to a project team in the planning phase. It will be very influential in determining what is needed to support seabed structures and the best options to protect them.

Seabed samples

Cores, grab samples and cone penetration tests are provide point data. However, there may be an unknown, difficult seabed between sampling points. It is very easy to assume that the seabed soils that lie between two sample points are homogeneous or that there is a linear change from one to the other. Taking seabed samples can be very time consuming and therefore costly in terms of vessel time. Despite this, any attempt to cut a seabed sampling programme should be resisted as the consequences of not knowing the nature of the seabed can have far reaching, embarrassing and expensive consequences.

Time and effort spent in geotechnical evaluation, including a thorough and painstaking desk study, is never wasted. Knowledge of seabed soil conditions can save frustration, time and money in the medium to long term and it should not be left to chance.

Rock berms

The rock structures are referred to as rock berms and as mentioned previously, they have a range of purposes. The prime purpose is to protect a pipeline, umbilical or cable from damage by trawls, anchors or dropped objects. In general, about 0.5 m of cover is just sufficient to provide protection. To ensure protection with a safety margin 0.7 m tends to be used. Representative diagrams of various types of rock berm are provided in Appendix F.

Specification

In considering the specification, the project team needs to be aware of what the primary and secondary functions of the rock berms, if any, are to be. Vertical tolerances and the top surface or top plane may need to be more precise for PLET supports and pads for jackup rigs. A precise tolerance is not possible with a large grade of rock such as 2 inch to 8 inch (50 mm to 200 mm). The project team and rock installation contractor should determine what is viable and the contractor's project team should suggest alternatives. Large size rock cannot produce a surface with a precise level tolerance; on the other hand small size rock will settle more and a greater tonnage may be needed for a given volume.

Rock berms can be specified by using a combination of the following parameters:

- Geodetic details for positioning and charting, spheroid, datum, projection and grid
- Type of rock; usually granite or schist (inert, igneous or high grade metamorphic)
- Size of rock; usually specified as an imperial or metric range e.g. 1 inch to 3 inch or 25 mm to 125 mm
- Tolerance related to the top cover e.g. –0.0 m to +0.4 m
- Depth of coverage above a reference such as the top of pipe
- Top width – usually a single value sufficient to provide protection
- Slope – typically 1 in 3 for a stable side slope, depending upon local conditions
- Diameter of rock berm for jackup rig legs or anti-scour berm around the base of a wind turbine tower
- Height above the top of the pipe or very occasionally the seabed
- Depth below the bottom of the pipe
- Line KP (to be used with caution because a grid position is better, where possible)
- Length of the berm in kilometres or metres between two defined points
- The extent of any counter fills
- Tonnes per metre, which is a very good method of providing protection for a product in a well defined narrow trench usually in stiff seabed soil. It will not work in softer soils where the rock will spread out to form a natural slope of 1 in 3 or thereabouts.

The depth of lowering and the trench depth are measured from the mean undisturbed seabed level. Any remedial rock installation would be calculated to provide sufficient cover to protect, and if necessary, stabilise the pipeline.

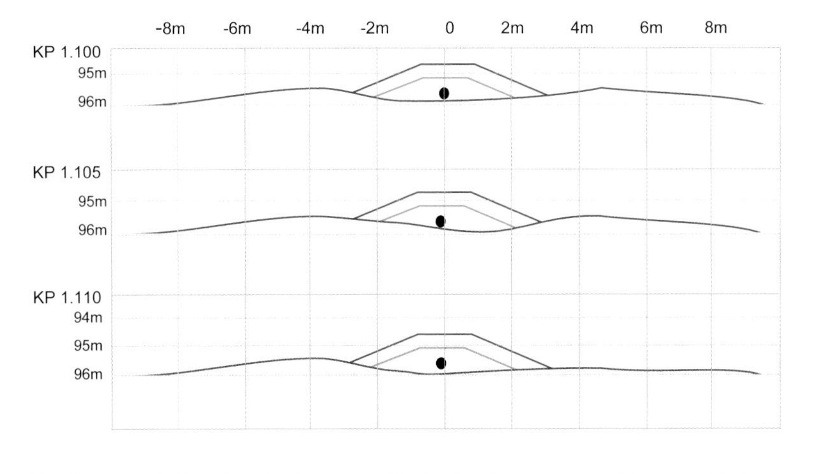

Figure 1 Representative cross profile design for a pipeline rock berm

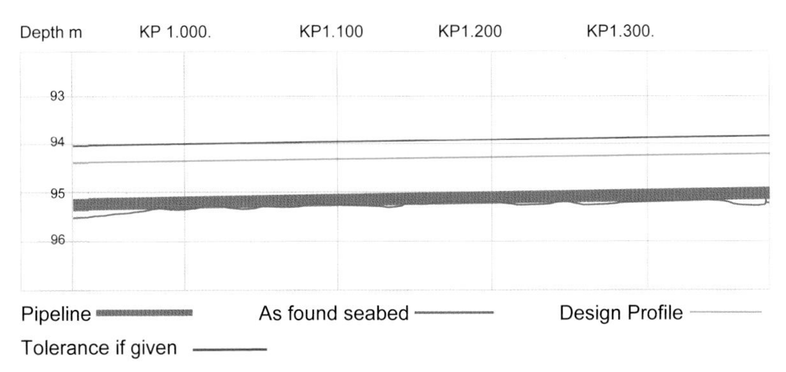

Figure 2 Representative longitudinal profile design for a pipeline

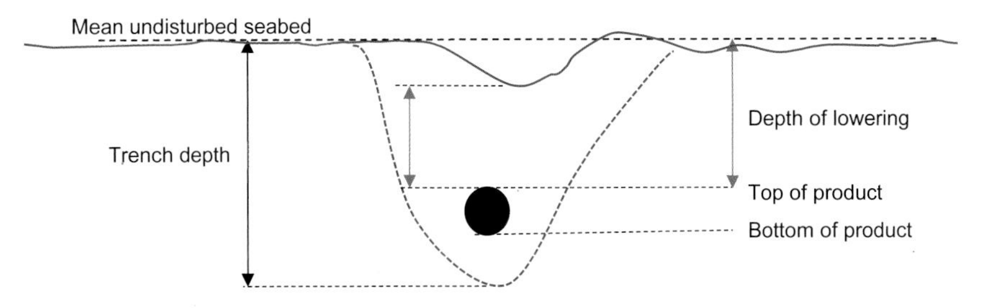

Figure 3 Representative levels used in SRI, trenching and ploughing operations

10 Quarries and quarry operations

Hard metamorphic and igneous rock underlies much of Norway, Sweden and NW Scotland. Aggregate exports are a growing industry in Norway. The Geological Survey of Norway (www.ngu.no) reports that the country exports in the region of 22.5 million Te of rock annually and around 6% of this is used by the offshore industry.

The features necessary for a quarry to service the offshore industry and exports are;

- preferably located close to the oil field
- a coastal location with access for deeper draught vessels
- a loading system capable of loading around 1500–2000 Te per hour
- a monitoring system for QA/QC
- access for service craft and or vehicles

Around 35 active coastal quarries are working in Norway and they are well distributed along the length of the coastline as far north as Hammerfest. A map showing the representative locations of 11 coastal quarries frequently used by the offshore industry is at Appendix D.

These quarries maintain a stockpile of the industry standard rock, which is currently 1 inch to 5 inch, and other well used grades. In the Netherlands sector the standard size is usually 1 inch to 3 inch. Some rock grades may need to be crushed and prepared to order in advance. A project with a high demand for rock may need to keep the supplying quarry well informed of its requirements in advance.

The coastal quarries in Norway have loading facilities consisting of:

- a single of multiple conveyor belt loading system
- an approach with enough water to enable a large deep draught vessel to berth and depart safely

General quarry view showing stockpiles and conveyor system

Rock crushing machinery at work

Quarry conveyor belt tunnel

DPFPV *Seahorse* loaded at Nord Fossen quarry

- a wharf capable of berthing a rock installation vessel
- a means of getting the appropriate size of rock from a stockpile to the loading system

Some quarries have an extending load arm to help in distributing the rock in the hold and loading belts can be moved along an arc or along the length of the wharf. Ships with excavators will use them to distribute the rock to create a stable load in the hold throughout loading.

Ships usually secure alongside at a quarry berth and they may not have to move along the berth to accommodate loading. However, more often than not the rock installation vessel will have to move along the wharf during the load out. At some quarries; notably Bremanger and Visnes, the ship's master may opt to keep the ship on DP and use the vessel DP system to maintain station during loading. It is normally permitted at the discretion of the master and the quarry management.

Before loading begins, the vessel staff will carry out a pre-loading draft survey to calculate the quantity of any residual rock and thus the amount that can be loaded. The quantity of fuel, oil, water ballast, etc. is known. Loading operations typically take between 12 and 24 hours, depending upon the quarry loading system, the ship and the quantity of rock required. Load rates are commonly in the region of 1500 Te per hour to a maximum of 4000 Te per hour. The final 10% to 15% of a full load can take disproportionally longer as the cargo is trimmed to avoid the risks to ship stability of a shifting cargo and vessel draught marks are also checked more frequently to maximise the load and keep within overall vessel weight limits. A ship's officer may directly supervise the last few hundred tonnes of the load and its distribution within the hold.

Tally hut display showing the hourly rate and cumulative load

Evening blast at Skipavika February 2019

Skipavika quarry, October 2015

During loading, a number of one cubic metre samples are taken from the stock pile and checked to ensure that the rock meets the specification for distribution of rock size within the load.

Rock is graded at the quarry by passing it through a series of grading grids as it is crushed. It must be stored with rock of similar grade and precautions must be taken to avoid contamination with larger gauge or even loose unprocessed rock in the quarry area.

The quarry sells rock by the tonne and monitors the quantities throughout loading. In some locations the load quantity is displayed in large numerals visible from the bridge. The vessel has a load limit and this is checked by a further post-loading draft survey, which will also accommodate any additional stores, bunkers, etc. The author has found that discrepancies are small and usually caused by rounding up or down. It is in both parties' interest to carry out very accurate monitoring and post-load checks.

Rock is detached from the quarry face by blasting. Quantities of explosive are strategically placed to break up a rock face into workable pieces. Once blasted, rock is processed to size by passing it through a series of heavy duty machines equipped with crushers to break up the rock and steel grids to grade the rock by size in order to achieve the grade required.

Quality control of the rock stockpile for offshore use is important. Neither contractor, client nor quarry management welcomes the risks presented by oversized rocks contaminating the load. An oversized rock is one where its longest axis exceeds 2.5 to 3.0x the maximum size of the chosen grade. In this case, the other two axes must be within the specified limit. For example, an elongated rock 12.4 inches long is within the size specification for 1 inch to 5 inch rock providing the other 2 axes are 5 inch or less. Using metric dimensions, a rock with a long axis of 310 mm is just in limits for 25 mm to 125mm rock.

It therefore follows that the gauge of rock being considered and smaller bore fall pipes must be matched to reduce the risk of a fall pipe blockage by rocks at the higher end of the specification.

Skipavika quarry from seaward, March 2019

Documentation at the quarry

The quarry will provide documentation to cover an analysis of the one cubic metre rock samples taken from the stockpile. It states a variety of parameters including the rock density, size range of the crushed rock, the dry bulk density of the crushed sample, the proportion of void space in the sample and it also shows a graph of the sample size. Quarry documentation is retained as part of the QA/QC for the contract.

In addition to the quarry sample check, ship staff will take a draught survey pre- and post-loading to calculate the quantity of rock taken onboard. The forward, midships and after draught marks are recorded and the calculation then works out the total displacement tonnage of the ship.

The following quantities are recorded: ballast water, and then consumables such as fuel, fresh water, water, lubricating oil and miscellaneous items. The total load is computed by starting with the difference in displacement between the pre- and post-loading draught surveys. The value of any consumables used such as fuel and lubricating oil is subtracted and the difference in ballast water is added to give the quantity of rock loaded. The weight of any residual rock from the previous load is then added to give a total load. Both the master and contractor's client representative will sign the draught survey and file it in the records.

Mixed load

Very occasionally a rock installation vessel will take a mixed load of two or more rock sizes. It is most important to keep the grades of rock separate, but also to minimise the loss of rock carrying capacity if possible. The only real option is to have separate holds for different grades as can be found in the *Stornes* and *Nordnes*.

In vessels where the hold is a large bin, it is possible to separate rocks of different sizes and grades by rigging a barrier in the hold. Alternatively each hold can be loaded with different rock grades. Backloading any unused rock may be an issue for the contractor.

11 Rock installation vessels

The word pretty is hardly ever going to describe a rock installation vessel. They have to be robust and highly functional vessels for the work they carry out. Distinct types of rock installation vessels have been developed and the first was the side stone dumper which is used to provide rock in shallow water during land reclamation and coastal engineering.

During the 1980s the fall pipe was developed for deep water operations in offshore oil and gas fields. In recent years, to meet the needs of renewable energy developments, multipurpose vessels have entered service. Multipurpose vessels can be used for coastal engineering projects in new and innovative ways and these vessels are equipped to carry out ploughing, trenching and cable laying, as well as rock installation using an array of methods.

Tideway Rollingstone loading at Jelsa Quarry in the Stavanger area, June 2013

Most fall pipe vessels have a top speed in the range of 10 to 14 knots, a faster vessel transit speed can be an attractive option to a client. Once an optimum speed is reached the power output, and therefore the quantity of fuel consumed, starts to increase rapidly. For example, a two to three knot increase in speed to 15 knots can increase fuel consumption in excess of 50%. Fast speeds can be achieved in benign weather conditions only. A vessel going into a rough sea at higher or full power risks damage. It also experiences the disadvantages of an uncomfortable transit and being slowed down by a heavy sea and strong wind. Additional disadvantages are a big fuel bill and the risk of damage.

Rock installation vessels have a large surface area and will be slowed down by a strong head wind. The author has been hove to whilst on a transit on a number of occasions. The worst case resulted in a 36 hour delay to what should have been a 48 hour transit on the Norwegian coast.

The author has also witnessed a high speed transit into rough sea. This resulted in damage to, and replacement of, many of the cable deck fittings. The bad weather also revealed a significant design fault in the ship's ventilation system. An exposed open ventilation trunk allowed sea water to enter the galley and discharge over the galley range. Locating the galley exhaust elsewhere or a simple cowl place over it would have avoided that problem. The consequences of cold sea water entering a hot fat fryer or coming into contact with live electrics are very dire indeed.

A ship is a big investment and it can take a number of years from design until entry into service. Once a ship is in service, it takes time for the whole system to work efficiently as personnel become more familiar with it and any hardware and control system issues to be resolved. Modern rock installation vessels have evolved from earlier designs as lessons have been learned and new developments and technology have been incorporated into the latest designs. They are complex vessels.

A rock installation vessel needs to be designed with:

- a rock storage and handling system to deliver rock to the area required
- thrusters at the bow and stern to make the vessel manoeuvrable and controllable
- a precise navigation and survey system
- control systems to keep the vessel in position and to deliver the right quantity of rock at the appropriate rate
- A ballast system to compensate for the rock being placed

Side stone dump vessel

The rock is loaded on deck and once the vessel is in position hydraulic rams

Aerial view of Van Oord side stone dump vessel *HAM 601*, placing rock around the base of a wind farm tower (© Van Oord and reproduced with their kind permission)

are used to push the rock over the side onto the target area. These vessels are typically small with a shallow draft to enable them to work close inshore at high tide. Side stone installation vessels commonly have a capacity in the range of 1500 to 3000 Te. The above photograph shows Van Oord's *HAM 601* working at a wind farm project. The four rock holds can be seen and rock is being installed from the port aft and the action of the ram can be seen in the gap between the two after holds. *HAM 601* and side stone vessels of its generation are highly manoeuvrable ships capable of operating in very shallow water.

Fall pipe vessels and flexible fall pipe vessels

The modern offshore rock installation vessels in service are large ships with complex machinery to deliver rock efficiently to the target, with integrated control and survey systems to enable rock to be accurately placed without the risk of causing damage to the pipeline, cable or any other subsea infrastructure.

Other than in shallow water, rock pushed over the side of a ship will spread out and a proportion of it will be wasted as it will inevitably fall outside

DPFPV *Seahorse* in transit to site with a full load, June 2012

Deck of DPFPV *Sandpiper* in June 2009 showing tracked excavator and grab like bucket

the target area. The fall pipe provides a controlled means of getting rock to the target area, in as efficient way as possible. For most operations, the fall pipe is usually set at a height of about six to twelve metres above the target and the rock leaves the fall pipe at a velocity in the region of two to three metres per second. Fall pipe bores range from 450 mm to 1000 mm.

DPFFPV *Tertnes* deck arrangement, October 2009 (Reproduced with kind permission of Van Oord)

DPFPV *Tideway Rollingstone* at Sandnessjoen, Norway in January 2011

The capacity of the first fall pipe vessels was less than 12000 Te, with the notable exception of *Sandpiper*, which carried around 17000 Te of useable rock. *Sandpiper* was converted from a large bulk carrier. The forward most and after most holds were both converted into DP machinery spaces and the remaining holds were used for rock. Its fore and after thrusters were placed where the vessel structure permitted but were not in the best locations. Because it was a vessel which had been converted to dynamic positioning it was not as manoeuvrable as those vessels built for the purpose with DP integrated into the design. Van Oord's early rock installation vessels *Rocky Giant* and *Tertnes* were also converted from bulk carriers.

A parallel development was the conversion of heavy lift vessels. *Tideway Rollingstone* was one of the first, followed by *Seahorse* in 2001. These vessels have all the superstructure forward and two holds, which can be subdivided for different rock types, mounted on the low freeboard broad beamed deck. The vessels have a modular fall pipe handling system and moon pool positioned between the rock holds that have similar capacities. The almost rectangular box-shaped hull form enabled retractable azimuth thrusters to be placed at the optimum positions to provide good manoeuvrability and positional control whilst on DP.

Newer vessels have been designed from the keel up as rock installation vessels, although their bulk carrier and heavy lift roots can still be seen.

Van Oord's rock installation vessels, *Nordnes* and *Stornes* are based on a bulk carrier design with the superstructure aft and a rock handling system based on hopper type holds. The vessels are designated DP flexible fall pipe vessels as the fall pipe configuration has more scope for movement than a fall pipe made of longer sections.

DPFFPV Nordnes approaching Kristiansund to load, September 2015 (Reproduced with kind permission of Van Oord)

DPFFPV *Nordnes* in the Bergen area, March 2019, the fall pipe stowage is forward of the moon pool area (Reproduced with kind permission of Van Oord)

Most vessels now have a means of placing rock over the side or stern, which is essential when working close to platforms and to provide scour protection for wind farm tower bases. The additional rock handling facilities will cause a reduction in the overall quantity of rock that can be carried. Some additional machinery can be removed for storage onshore when not required. The simple forms of an inclined fall pipe or a chute loaded by an excavator are very useful. Other means such as extended conveyor belts were developed to enable rock to be placed over the side or over the stern of the vessel.

DPFPV *Simon Stevin* inclined fall pipe on deck (Reproduced with kind permission of Jan De Nul Group)

Rock installation close to a platform via a side chute (© Mitch Foster and reproduced with his kind permission)

DPFPV *Flintstone* at Aquarock Quarry Sandnessjoen, May 2012

DPFPV *Rockpiper* in transit from Jelsa Quarry with a full load, September 2018

Multipurpose vessels

Multipurpose vessels have been developed to service the growing offshore wind farm industry. Designers have combined a flexible rock installation capability with ploughing, trenching or cable laying. The vessels are medium sized and primarily designed for work in shallower and coastal waters. The downside of a multipurpose vessel is that by trying to accommodate many functions, it also requires compromises within configuring project machinery

Van Oord multipurpose vessel *Bravenes* (© Mr Barry Van der Meijden and reproduced with his kind permission)

and project personnel skill mix. By good design and the right mix of skills on board they have shown themselves to be very versatile vessels in coastal engineering projects. The mindset is that it is an integrated flexible vessel like many construction vessels already servicing the offshore industry.

Rock delivery system

The rock delivery system comprises the holds, rock handling system, fall pipe and fall pipe ROV. This is complex machinery with multiple moving parts, telemetry and video monitoring systems. In some vessels the area enclosing the moon pool, fall pipe stowages and associated machinery and control rooms is referred to as the module.

For all intent and purposes, the fall pipe can deliver rock within the footprint of the moon pool or module area depending upon how much scope there is to move it. A flexible fall pipe comprised of shorter sections will have a greater scope for movement. Modern applications require rock to be placed outside of the vessel hull footprint; for example, when working close to a platform and when building anti-scour berms around wind farm towers. Of necessity, the rock installation contractors have become adept problem solvers.

FFPV *Stornes* deck and tower with hatch covers closed at sea

Companies have developed solutions including multipurpose vessels, which will no doubt be refined and complemented in the future. The methods include the use of an inclined fall pipe, a chute with free fall rock and extended conveyor belts over the side or stern of the vessel. Each of these methods will require some modification to the handling system and more weight on deck will result in a slight reduction in rock carrying capacity

The hold

Most ships will have a number of holds. The forward and after holds are usually separated by the moon pool and fall pipe module area. There may be a difference in capacity between the forward and after holds as is the case with *Stornes* and *Nordnes*. In these vessels the holds are separated by transverse bulkheads and covered by hatch covers when at sea. Each vessel will have its own designations for each hold.

The holds are shaped to enable the excavators to extract virtually all the rock or to allow all but a residual quantity to fall by gravity to the conveyor belts. *Sandpiper*, which was one of the earliest rock installation vessels, nearly always retained in the region of 1000Te on board, which for all intents and purposes was inaccessible.

The photographs of FFPV *Nordnes* and FFPV *Simon Stevin* show the holds at an early stage of loading.

Vessel hold showing aperture to the conveyor belt (Reproduced with kind permission of Van Oord)

DPFPV *Simon Stevin* forward hold empty and ready for loading (Reproduced with kind permission of Jan De Nul Group)

Rock handling system

The rock handling system is designed to deliver rock from the hold to the seabed via a conveyor system to the fall pipe. Rock can be unloaded from the holds from above with an excavator shovel or grab into a hopper, or through an aperture at the bottom. The rock is then transported on a conveyor belt to a hopper close to a conveyor belt feeding the fall pipe. Fall pipe chutes channel the rock to ensure that it goes through the pipe and not outside it. Fall pipe chutes may be a simple open funnel shaped device or more elaborate covered assemblies.

On *Nordnes* and *Stornes* the rock is moved from below the holds up over the forecastle by a conveyor belt system and onto a conveyor belt mounted on a boom, which is placed above the forward hatch covers. The conveyor belt arm can be rotated and raised to clear the hatch covers for loading. The boom can also be used to discharge the cargo to shore or another vessel, if the ships adopt their secondary role as bulk carriers. The fall pipe is made of short stackable plastic tubes linked by chain and stored horizontally in purpose built stowages when not in use.

Fall pipe ROV

A fall pipe ROV (FPROV) differs considerably from a free flying or tethered work class, survey or observation ROV. They are large, weighing from between

Vessel conveyor belt system from aft looking forward

DPFPV *Simon Stevin* after conveyor belt (Reproduced with kind permission of Jan De Nul Group)

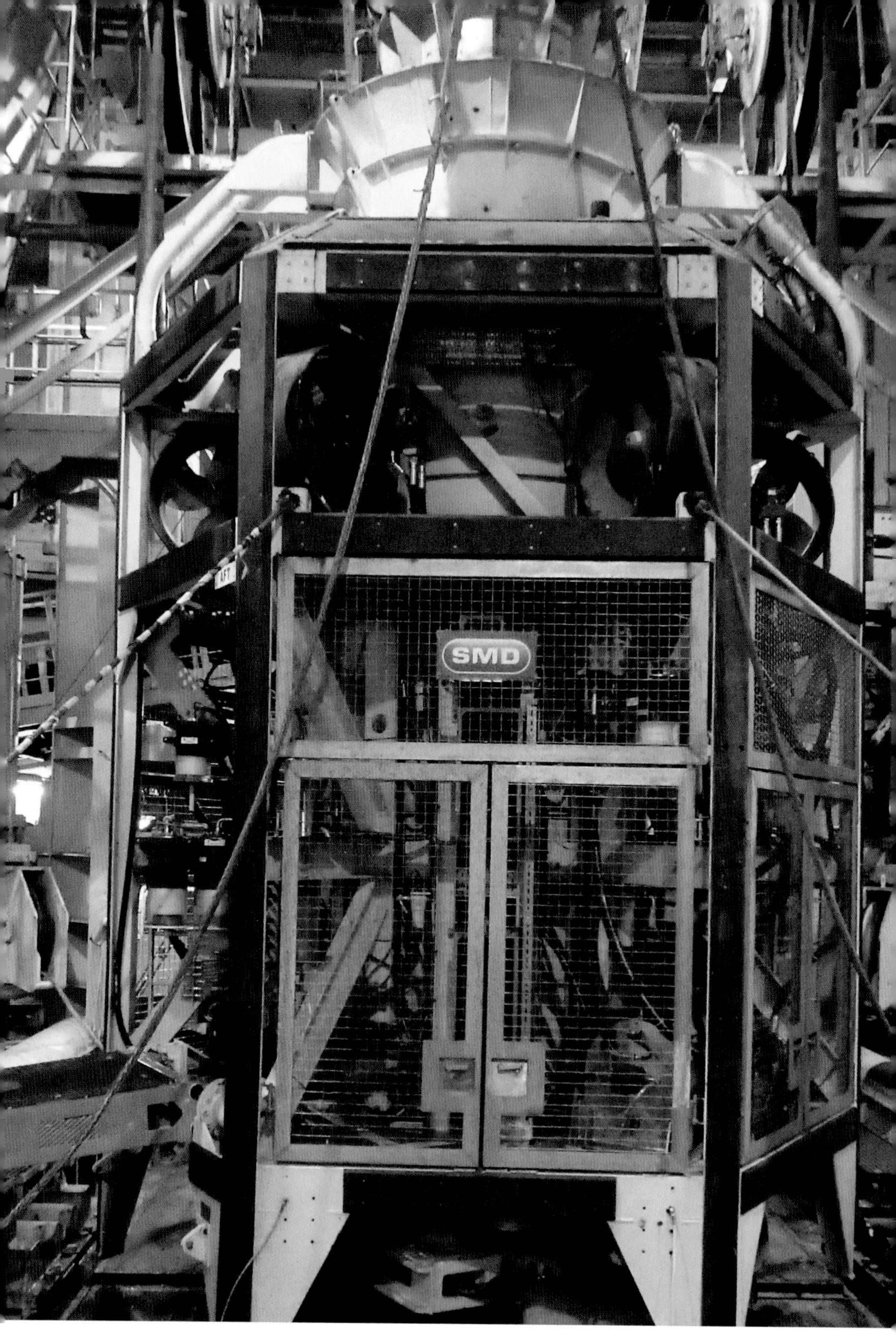

DPFPV *Simon Stevin* FPROV on deck (Reproduced with kind permission of Jan De Nul Group)

7–10 tonnes to in excess of 30 tonnes. It is built around a cylindrical tube to accommodate the final section of fall pipe. Each fall pipe ROV is different, and although industry standard components and instruments are usually fitted, they range in construction design from a conical shape to a steel framework or hexagonal. To minimise rock wastage and improve efficiency, each company has developed FPROV attachments to direct the flow of rock leaving the fall pipe. These devices are particularly useful when working around structures to prevent rock getting into them and risking damage, or when building berms over glass reinforced plastic (GRP) covers to hold them down on the seabed and to reduce the risk of trawl damage.

A fall pipe ROV lacks a manipulator, although some have hydraulic arms from which an acoustic DP beacon can be launched or recovered. Deploying a beacon with the FPROV is relatively straightforward, although its recovery can be problematical because the operators are relying on camera views and they have no depth of field. Recovery is further complicated by the size and manoeuvrability of an FPROV.

Like a free flying ROV, the fall pipe ROV has thrusters, lights, cameras and instrumentation. It is designed specifically to facilitate rock installation operations only. FPROVs are large, heavy and lack a free flying ROV's manoeuvrability or manipulators. They have full vertical movement but lateral movement is severely restricted, especially when the fall pipe is deployed. Because FPROVs lack the manoeuvrability and the camera configuration of

DPFPV *Simon Stevin* FPROV winch (1 of 2) (Reproduced with kind permission of Jan De Nul Group)

a WROV or an observation ROV, they are therefore not designed for visual survey works. An FPROV can be lowered, raised and even recovered to deck with the fall pipe in situ. Its lift arrangements are separate from those of the fall pipe. In some vessels, the fall pipe and FPROV winches can be synchronised to allow the operator to use a single control for fine adjustment.

The FPROV can be agile in vertical movement as the main lift umbilical is used, but is restricted in lateral movement to the natural scope of the main lift and even more restricted to the limited scope of the fall pipe when it is in place. The fall pipe scope for lateral movement will increase with the pipe length, and hence depth, of the fall pipe and will take into account any risk of overstressing.

The FPROV is launched vertically into the moon pool; the main lift and umbilical are usually combined. Like a conventional ROV, they carry instrumentation such as acoustic beacons, MBES transducers, bathymetric depth sensor, altimeter, sonar and cameras. At the time of writing only *Bravenes* and *Livingstone* have a free flying work class or observation ROV, although one was fitted in *Sandpiper* and a WROV can be readily fitted in both *Simon Stevin* and *Joseph Plateau*. A free flying ROV can be operated close to the seabed and can be fitted with a pipe tracker if required. However, free flying ROVs are costly to mobilise to a ship and, in all likelihood, would be a nice to have, rather than an essential fitting.

Some clients may specify a pre- and post-installation visual survey, which can realistically only be achieved with the use of a free flying ROV fitted with MBES using two transducers in profiler mode, bathymetric depth sensor, altimeter, centre line and boom cameras. In the author's experience there is little benefit to having a free flying ROV on board a rock installation vessel. It could even come with extra costs with the requirement of additional project crew and resources. The skills required to operate a fall pipe ROV are different from those required to operate work class or observation ROVs, although there are areas of commonality.

An ROV which is used infrequently and moved to and from different locations, is susceptible to defects. Even when placed in a hangar, defects arising from a lack of use may include condensation damage, short circuit damage and seal failure.

In a forthcoming vessel conversion the fall pipe remotely operated vehicle is being called a fall pipe remotely operated tool (FPROT). It is a more succinct description and will not convey the wrong perception to client company personnel.

Fall pipe system

The fall pipe system includes the machinery needed to deploy it and to hold it in place. There are three methods currently in service and the majority of

DPFPV *Simon Stevin* aft fall pipe winch (Reproduced with kind permision of Jan De Nul Group)

vessels use a system of heavy guide wires and pins or hooks to hold the fall pipe in place. A specialist crane is used to move pipes from their stowage to the moon pool. This system has heavy duty winches and gallows to hold the pipe in place. *Flintstone*'s fall pipe uses a bayonet connection to join the pipes and a Huisman handling system supports the fall pipe. The latter is unique to *Flintstone* and unlike in other vessels the fall pipe sections are stored horizontally. *Stornes* and *Nordnes* use a chain system to handle the buckets on deck and when deployed.

Fall pipe systems have a water inlet to mix sea water with the rock. The sea water acts as a lubricant and reduces the density of material in the fall pipe. It also prevents the build up of a potentially damaging pressure differential in the pipe.

The winches and cables are heavy duty to enable the fall pipe to be deployed at depths down to 2000 m. They are built with a safety margin to accommodate additional loads in the fall pipe arising from a blockage. Fall pipes are heavy, robust and yet precision components in the rock placement system. They are made of steel, aluminium, heavy duty thick plastic or glass reinforced plastic (GRP) sections, which are secured and depending upon the system fitted are held in place by a combination of main lift guide wires, pins, chains, hooks or a bayonet arrangement.

DPFPV *Simon Stevin* during SRI operations (Reproduced with kind permission of Jan De Nul Group)

Newer systems are designed to minimise handling, but do require monitoring by closed circuit television (CCTV) with an operator observing the process. The person operating the control system accepts each stage of the process following a positive report from the observer. In older systems the fall pipe sections are pinned together by hand under the close direction of either a bosun or deck engineer.

Fall pipe sections are typically 8–12 m in length and internal bore diameters vary from 450 mm to 1000 mm. Steel fall pipe sections may be lined with an industrial grade neoprene liner to avoid mechanical wear caused by rocks coming into contact with the pipe. Steel pipes come in various thicknesses, the tendency is to use thicker heavier steel pipes close to the surface; and in deep water thicker gauge steel pipes are used close to the FPROV. Steel pipes are susceptible to corrosion and good preservation and maintenance is necessary to optimise pipe section life.

Nordnes and *Stornes* are the exceptions in being fitted with a very flexible fall pipe, the majority of which is formed from a chained stacking plastic bucket system consisting of two-metre sections with a bore of 1000 mm.

Hazards to the fall pipe

The fall pipe is essentially a simple concept of placing rock down a tube. In addition to the weight of the fall pipe, there is some frictional force between the descending rock and water mix and the side of the fall pipe. There is a safety margin to cope with a volume of rock forming a blockage and overload alarms to alert the operators to it. However, despite its conceptual simplicity, there are a number of hazards to the fall pipe including blockage, failure and buckling. The fall pipe is managed and monitored to maintain optimum operating conditions.

Large volumes of rock in contact with or striking the fall pipe can eventually damage or erode it. To overcome this, fall pipe sections are not kept in the same vertical position in the fall pipe for each use. Neoprene liners are colour layered and any wear can be readily identified by observing a change in colour.

Fall pipe blockage can be mitigated by good quality control at the quarry which first of all must avoid contaminating rock stockpiles with oversized rocks. The quarry staff will monitor the distribution of rock sizes from representative samples of each load, to ensure an optimum size distribution. Blockage by rocks can be caused by a number of oversized or close to oversize rocks descending together and locking to create a barrier. Alternatively, fine material bonding together in concretions can cause a blockage. It may also be possible for a blockage to be caused by a neoprene liner detaching from the inside of the fall pipe.

The risk of a fall pipe failure can be mitigated, but not eliminated, by a planned maintenance schedule and by rotating individual pipes to avoid uneven stress in the pipe section. To the untrained eye most fall pipe sections look alike. However, heavier gauge steel fall pipes are used at the top and at depth for deeper rock installation. Each section is identified by a serial number and a log is kept of individual use which includes details of its position in the whole assembly and the time deployed, so that a maintenance routine based on wear can be scheduled. Fall pipe sections are landed for refurbishment onshore in accordance with a planned maintenance routine.

As well as having the potential to cause damage to subsea assets, the fall pipe and FPROV can be damaged by impact with a subsea structure, the seabed or anchor wires. Any compromise of anchor wires is always thoroughly investigated and the wire is checked for any damage. Damage to anchor wires, the fall pipe or any structure can lead to the loss of productive time and the additional costs associated with investigating and then repairing any damage.

It is therefore of vital importance that the client company provides full details of the field layout and bathymetry for the area of operations. Once in the field all risks of collision are to be mitigated by careful planning to keep a safe distance from and a safe height above any hazards. When operating in confined areas all personnel must show a heightened sense of spatial awareness.

Extreme and catastrophic fall pipe failures resulting in the requirement to recover sections of the fall pipe from the seabed are extremely rare. Fall pipe failure can be attributable to material fatigue, incorrect use or lack of use of the water inlet pipe, system or human error in failing to secure the fall pipe properly. Any fall pipe failure is the subject of a thorough investigation and measures will be put in place to prevent its reoccurrence.

Night time operations close to a platform

Fall pipe buckles caused by fatigue or stress are another albeit rare potential hazard. Vessel speed is restricted to about 1800 m/hr with the fall pipe deployed to avoid the risk of buckling.

Shorter in-field transits with the fall pipe in situ are made to reduce the time spent in launch and recovery when compared with a normal transit. The vessel superintendent and master will compute the maximum sailing distance between berm locations with the fall pipe fully or partially deployed. It follows that at greater distances between rock berms, it is more cost effective to recover the fall pipe before departure and to redeploy it on arrival. Additional time may be spent to survey a corridor between SRI locations in a work area. A transit with the fall pipe down will not be attempted without detailed knowledge of the seabed bathymetry in a corridor between the points. As a general rule, the route will be as direct as practical taking subsea structures into account.

12 Operational considerations

A successful offshore phase is the result of careful preparation and execution by a highly trained and skilled team both onshore and offshore. This is backed up by good detailed planning and dialogue between the client and the rock installation contractor. Their dialogue should cover all aspects of the project from initial planning and specification, to the conduct of operations and final deliverables.

Organisation onshore

Before the project is approved for offshore work, considerable effort and dialogue will have taken place between the client and contractor. Onshore organisation would normally be delegated to client and contractor project managers or specialist project engineers and surveyors. They have day-to-day management for the project and provide a point of contact when the work is in progress offshore.

Logistics

Logistics are a most important aspect of any operation. Rock needs to be ordered, and a quarry berth booked because congestion and delays at quarry berths can occur. Norway has a choice of well equipped quarries. In some parts of the world rock has to be transported to the vessel and occasionally a temporary loading system has to established. *Simon Stevin* carries a loading system that can be set up in port to enable loading where there are no facilities.

Arrangements must be in place for delivery of stores, personnel changes, and disposal of general waste and waste oil. The ship must have enough fuel to comply with regulation and company marine procedures. Fresh food has to be managed carefully to avoid waste. Stores, spare equipment and personnel must be in the right place at the right time and prevarication inevitably leads

A stores tender back loading waste for disposal

to the loss of options. Suppliers may choose to back charge for any additional costs incurred by late changes of plan.

Logistic arrangements for a rock installation vessel are, on the face of it, much easier than a construction or survey vessel that may be continuously on site for a number of weeks. A single rock installation trip is rarely more than ten days at sea, including any weather standby or breakdown. The regular turn around at quarries gives the project staff regular opportunities to arrange fuel and stores and to mobilise and demobilise personnel and equipment. However notice is often short as is the ship's time in harbour.

Catering staff will manage loading of provisions carefully, a longer than anticipated trip and future provision replenishment can lead to a shortfall in provisions. Good catering is an essential feature for welfare and good morale onboard any ship.

Ship's agent

Ships' agents are an integral part of a successful port visit. They are highly experienced in ensuring that short port visits to reload are run smoothly with simultaneous activities taking place whenever practicable. Ship's staff and visitors will normally arrange transport to and from an airport or hotel through the ship's agent. Agents are exceptionally well organised for moving personnel to and from the vessel, and they do react efficiently to accommodate short notice personnel changes. However, they are not magicians and good liaison is a must.

A stores tender ready to unload

DPFFPV *Stornes* refuelling at Skipavika quarry

Fuel

Outside of the Bergen and Stavanger area, where mobile fuel barges operate, it may be necessary for the ship to take fuel and stores at a berth remote from the quarry. In the worst case, particularly in smaller centres, fuel may have to come by road in fuel bowsers. This can extend a port visit by up to eight hours.

Personnel

Most companies moving personnel will use a travel agent or have an internal travel department. It is useful to know that air travel to smaller towns in Norway is often limited from early Saturday until later on Sunday. Summer timetables can also reduce the number or change the timing of flights in Norway, especially to and from some smaller towns. It is important to get personnel travel arrangements in place in a timely manner. This is important, but especially so if any weekend air travel to smaller towns, long road journeys, ferry crossings or boat journeys are involved.

During wind farm projects some personnel may be moved to and from the ship by crew transfer vessel (CTV). Those being transferred within a wind farm complex must have the appropriate training, certification and PPE. The CTVs must be certificated and boat landing ladders equipped with fall arrestors must be used; the use of pilot ladders, other than by pilots, is prohibited. Boat transfers can take place in sheltered waters, but are usually strictly forbidden during oil and gas projects. In Norway frequent use is made of water taxis. Vessels working on nearshore projects like wind farms can be distracted by regular CTV transfers for day visitors.

Routine personnel transfers by helicopter to and from rock installation vessels fitted with a helicopter deck are unusual. Helicopter deck operating regulations and operating limits are very strict, and at the time of writing (January 2019) only *Simon Stevin*, *Joseph Plateau* and *Livingstone* have a helicopter deck.

Contractors will need to source rock from a quarry within reasonable reach of the work site; naturally price and availability of stock are prime considerations. The quarries at Dirdal and Jelsa are near Stavanger, those at Slovag and Skipavika are near Bergen. Both major cities are well served by airlines. Averøy and Visnes quarries are near Kristiansund or Molde which have limited flights on a Saturday and Sunday. The quarry at Nord Fossen is a few hours' drive from Trondheim, Norway's second city. Aquarock quarry is close to Sandnessjoen, which is serviced by only one flight on Saturday and two on a Sunday. The quarry at Bremanger is a short boat journey from Florø, which again has limited air services over a weekend. A map showing the locations of eleven quarries on the Norwegian coast is provided in Appendix D.

Approaching a bulk carrier to reload

Loading from a bulk carrier whilst on dynamic positioning

Practically the whole of Norway closes down for Norwegian National Day on 17th May and so it is best to be at sea before then. Norway also has strict pilotage and vessel traffic requirements, especially around the busier ports. Pilotage is usually compulsory for foreign registered vessels. The masters of Norwegian registered vessels can be exempt from taking a pilot, if they have fulfilled local knowledge conditions. Pilot availability can cause delays, especially in the summer time when cruise ships are sailing the Norwegian coast. There are currently no Norwegian registered rock installation vessels. The author has witnessed self pilotage in Norwegian waters only during a period of industrial action.

To reduce time sailing to and from a quarry, some contractors may use bulk carries with a cargo discharge capability at a harbour closer to the work site. In 2013 during extensive SRI works for Total E & P's Laggan Tormore development, Tideway's vessels *Flintstone* and *Tideway Rollingstone* were loaded from a bulk carrier berthed at Dales Voe, north of Lerwick, which saved around two days in transit time per vessel per load.

Communications

The art of communication is that the originator of a message must be confident that the intended recipients have received it and more importantly understood it in the way the originator intended. With the plethora of devices and media and with the apparent ease of communication today, it is somewhat ironic that communications have not improved.

Radio communications at sea have developed beyond recognition in the past 25 years or so. Modern maritime communications are now largely based upon the VSAT (*V*ery *S*mall *A*perture *T*erminal) satellite system, which has largely replaced the HF and VHF coastal radio stations. The VSAT system is any two-way internet satellite system that has an antenna smaller than three metres. Marine VSATs are kept stable by gyroscopes so that however the vessel moves, the satellite in question is kept in line.

Verbal and written communication

Verbal and written communication must be unambiguous and the message needs to be delivered to the right people via the management chain. It is easy to discuss an issue at a management meeting then fail to cascade it to the people doing the job. Shift meetings, toolbox talks and time out for safety (TOFS) should be used to keep all personnel informed of any work-related issues concerning them.

Radio communication

Operating HF radios is an art and the master and radio officer used to control all outgoing communication. The author recalls the difficulty, expense and poor quality of HF radio telephone calls. The frequencies were invariably single side band which distorted the voice and the signal was further degraded by static interference. Coastal radio stations would allocate ships a slot and it was always hoped that it would not be behind a large cruise ship.

Coastal VHF radio stations were slightly better, but again the ship was in a stack until an operator became available. Confidential issues could not be discussed as the conversation could be heard by others trying to contact the radio station. VHF remains the prime voice communication between vessels and other marine users.

Telex was also available. Telex worked on HF but could be temperamental and a knowledge of HF radio propagation was useful. It was possible to see a coastal radio station yet fail to get an HF signal. Many hours could be spent trying to find a coastal radio station and suitable frequency to clear telex messages.

Satellite communication

VSAT and digital communications spare offshore and project personnel from those frustrations associated with using analogue HF and VHF communication via coastal radio stations.

Most ships will also be fitted with an alternative INMARSAT satellite communication system which is usually only available to the master and bridge team. All rock installation vessels have a number of extensions of the company's main telecom network linked into main networks via VSAT. Telephone and email have become the prime means of communication at sea, although some caution should be exercised before sending large files to a ship via email.

Voice links offshore will rarely, if ever, be as good as those onshore. There is usually a short but noticeable signal delay and some loss of signal quality. Satellite bandwidth can be an issue resulting in slow data rates, although most contractors will set aside some bandwidth for the sole use of the clients. Satellites tend to be stationed above the Earth's tropical regions. This can be problematic when operating in higher latitudes where the satellite elevation is low and the line of sight to the satellite from the ship may be blocked by adjacent topography particularly in Norway, platforms, drilling rigs, FPSOs and a ship's structure.

Satellite receiver domes are located as high as possible in a ship; the downside is that a vessel VSAT system may have difficulty holding a signal if there is too much vessel movement or vibration.

There is a considerable, though unquantifiable, domestic benefit in having universal internet communication in a ship. Good communication is an important beneficial welfare facility. Asian crew members may be on board for up to six months without the opportunity for a break for shore leave. Easy contact with home is therefore of vital importance.

Project staff onshore may also find occasional communication outages very frustrating. However, the author can reassure readers that communication facilities are infinitely superior to those available up to the mid-1990s.

Social media

Although not directly related to a project, the topic of social media has to be addressed in outline. There are many positive aspects to social media, especially in helping people keep in touch with friends and family. With a greater and broader access to rapid communication, it does have a big downside, in that unguarded comments can betray confidentiality which has to exist between a client, the main contractor and any subcontractors.

Many companies have an expectation of confidentiality which is either specified or implied. Personnel who work on offshore projects must be aware of client confidentiality and comply with it. Whistle blowing, should it prove necessary, has to be made to the relevant authority and not through the mainstream or social media.

All companies have a system for reporting safety issues and grievances and they are taken seriously. Each ship has a Designated Person Ashore (DPA) whose contact details are displayed on board and all personnel have access to the DPA if they wish. An open public forum is not the place to express such issues because it clearly may place the client-contractor relationship at risk and a real grievance will be overshadowed by the loss of trust. In the UK the libel law is applicable to open social media with both the provider and individual having a responsibility.

Organisation at sea

All ships have a designated minimum safe manning crew. When working on a project the vessel is organised with a marine crew and project team working together for project delivery. For a project, the vessel crew is enhanced above the minimum safe manning level with additional engineers and DPOs to provide the requisite cover for working in DP modes. Also, additional catering staff and deck personnel may be added to the crew.

Each rock installation contractor's team composition and skill mix will differ slightly, with each company having a slightly different way of organising the crew offshore.

The master and offshore manager or superintendent work closely together to manage the safe and efficient execution of the project. The master always retains responsibility for the safety of the ship and all personnel onboard. The superintendent or offshore manager is responsible for the technical planning and execution of the project offshore.

Management

The on-board team may have a number of jobs in progress such that reporting of previous contracts will be at an advanced stage, a job will be in progress and preparations for the next contract will be underway. Any breaks for bad weather or defects enable the on-board project team to progress processing and reporting. When a project team is demobilised the survey processors and offshore works manager stay on board to complete any reporting before despatching it for office QA, QC and submission to the client.

Most vessels will hold a vessel morning meeting to review safety issues, progress, discuss planning and logistics. Key vessel, project and client personnel will participate in a project conference call daily. If there are any issues or difficulties arising, the project managers onshore should be made aware before

DPFPV *Nordnes* operating close to a platform (© Van Oord and reproduced with their kind permission)

the conference call takes place. It is reasonable that any contentious issues are brought to the attention of the project staff onshore at the earliest opportunity.

Each contractor's organisation on board is slightly different. Good communication is vital as time is often short. The master has to make a number of arrangements with the ship's agent, clients and contractors. There is a short period for stores delivery, fuel and personnel movements. The quarry management need time to organise berthing and loading where workers are subcontracted. Port authorities require notice for pilotage, which is more acute during the summer months when the Norwegian coast is busy with cruise ships and it is a traditional Norwegian holiday time.

Bridge personnel

The DPOs, ROV pilots, fall pipe operators and survey team carry out the work as directed by the master and superintendent. Some contractors combine the roles of DPO and ROV pilot, whilst others keep the roles separate. Sufficient numbers of personnel are made available to ensure a rotation during the shift to offline and vessel duties. The senior watchkeepers will almost invariably have a ship master's certificate of competency and plenty of sea going experience. The bridge is designed for efficient work flow communication between the DPOs, survey and rock installation teams.

Simon Stevin Bridge from the starboard side (Reproduced with kind permission of Jan De Nul Group)

Simon Stevin fall pipe operator's position (Reproduced with kind permission of Jan De Nul Group)

Rock installation team

A superintendent or supervisor will oversee the rock installation process during each shift. Job titles vary with company and fall pipe operators or FPROV pilots may be multiskilled and their duties may be combined with DPOs or deck engineering.

This team comprises the maintainers and operators including excavator operators. The fall pipe operators or FPROV pilots run and monitor the rock installation process, from building the fall pipe ROV through to operating it and unrigging it. As well as steering the fall pipe ROV during operations, the bridge team controls the rate of rock deposited, monitors the control system to provide rock to the fall pipe and the CCTV covering the system. They react to any emergency involving the system and work closely with the DPOs and survey personnel to make sure that the fall pipe is tracking above the location where rock is required. This team is pivotal in effectively building the rock berm. Remedial works are a fact of life but a good team will reduce them to a minimum.

The rock installation system is maintained by the deck or fall pipe engineering team. Rock installation is heavy work and the equipment is very robust to withstand the stress of the operation. Despite rigorous planned maintenance routines, system components break down through wear and tear and shake loose with vibration. Exposed electronic components can be damaged by water ingress, vibration, heat and condensation. Highly skilled personnel are needed to monitor the system, and to carry out planned maintenance and any repairs.

Survey personnel

The survey team, under the direction of the survey party chief, monitors progress, logs data and provide users with positioning information. The surveyor on shift is referred to as the online surveyor. The online surveyor will be highly proficient in the correct calibration and operation of the survey equipment. The highly important data management function is also delegated to the survey team. The survey team is completed with offline surveyors also known as data processors. The offline complement may be staffed with personnel with a surveying background; others may have an industrial or cartographic drawing office background. Geographic information systems (GIS) graduates are also working offshore as data processors. Their function is fundamental, as it is a verification of the work.

The data processing team provides graphical survey data to plan, monitor progress and to confirm that the work is complete. They also ensure that raw data is properly stored for future reference. Several contractors are moving from paper records to electronic records. The author is of the opinion that paper records still have a place and that the option of using paper needs to be available for the time being.

The online survey position on DPFPV *Simon Stevin* (Reproduced with kind permission of Jan De Nul Group)

DPO position adjacent to the fall pipe operators on DPFPV *Simon Stevin* (Reproduced with kind permission of Jan De Nul Group)

Client representatives offshore

Client companies will appoint an offshore representative to monitor progress and to act as a liaison between the project staff onshore and the contractor. They are the eyes and ears of their respective companies. Some companies may appoint a number of representatives to cover 24 hours or to manage different disciplines. Contractual and insurance requirements do not often allow contractor self-certification. If the rock installation is subcontracted, both the main contractor and field operator will have representation onboard.

The operator's representative has the prime function of liaising with personnel on any platforms, drilling rigs or drill ships operating in the field and ensuring that the correct permits are in place. The contractor's representative has the more high profile role and acts as a liaison between the contractor's project staff onshore, the operator's representative onboard and the subcontractor's team on board. The first day is often spent clarifying any final detail regarding the job. The contractor's representative also scrutinises and accepts the pre-SRI designs and the finished product and will present the accepted data to the operator's representative for review. The deposit consent (DEPCON) should be monitored by a client representative to keep project staff up to date. This is relevant where further stabilisation is anticipated and there is no spare capacity built into the DEPCON.

In order to ensure that the client has a good SRI phase it is very useful to appoint representatives with knowledge and experience of rock installation. The author has seen social science graduates on their first ever trip to sea act as operator company representatives during third party works. Fortunately, the third party contractor had a proficient representative onboard.

The client representatives should develop a working relationship with each other and ship's staff. They can act as facilitators for any problem solving. Any contentious issues that cannot be satisfactorily resolved on board should be referred to the respective project personnel onshore.

Permit to work

The field operator will give details of any permit to work requirements for rock installation, which will vary from project to project. Permits can be set up in advance of the vessel's arrival infield. In 500 m safety zones for platforms, FPSOs, and drilling rigs, a permit is always required. Work in open water sites and pipeline systems outside a 500 m safety zone may also be controlled by permit. It is always prudent for the operator's offshore representative to contact the area authority to establish communication in advance of the vessel's arrival on site to ensure that the permit system is managed to comply with the operator's requirements. Platforms and drilling units have specific times for permit issue, usually at 07:00 and 19:00; the control room personnel issuing them tend to be very busy around these times and patience is necessary.

Pre-planning

The superintendent and survey party chief will usually have some time to make outline preparations for the project in advance of the pre-installation surveys. They will work out the rock placement plan that includes the order of work, the route and start and end positions. A long berm will be progressed in segments to facilitate client approval, avoiding long tracking runs, and reducing the number of any fall pipe changes. Experienced works managers or superintendents may test their plan in small lengths to find a means of reducing the prospect of any time consuming remedial works.

Survey

Survey data is gathered by combining position derived from GNSS and acoustic positioning with MBES data. MBES forms a swathe of data which can be processed to give a spot depth in gridded squares which for SRI is usually a grid of 20 cm squares. After correcting for motion and any tidal differences, the data is formed into a digital terrain model (DTM). The data can be displayed in a variety of ways such as a contoured plan view, a colour-banded plan or oblique view, long profiles and cross profiles. With the use of accurate positioning systems and large scale charting, fine detail can be shown at horizontal scales as large as 1:500.

On-site checks

On arrival on site the following procedures must be carried out; DP checks, position checks and MBES calibration and any further calibrations arising from system checks.

Vessel operating limits

All vessels have operating limits for SRI. These limits comprise a complex matrix of wind, sea state and vessel motion, especially during critical phases such as the launch and recovery of the FPROV and fall pipe. Where fall pipe sections are manipulated by a specialised gantry crane, the crane operating limits will determine operations. It should be noted that the fall pipe handling systems are stowed on the port or starboard side of the moon pool or module area, high above the vessel's pivot point and will be subject to most vessel movement. The fall pipe handling gantry crane can move to any pipe location on the pipe deck. Masters will veto equipment launches if imminent bad weather is approaching.

Both *Nordnes* and *Stornes* have chains of two-metre stacking plastic fall pipe sections. These are manoeuvred along the horizontal handling rails and then over the tower into the moon pool. A chain of buckets can be configured to make a pipe of the desired length, which gives both vessels a wide operating envelope.

There is some scope to steer a fall pipe laterally using the FPROV thrusters, which for a fall pipe of 150 m length, is typically around a five metre radius. The radius will increase with the length of fall pipe. Flexible fall pipe designs allow more scope for lateral movement than rigid fall pipes.

DP checks

Dynamic positioning (DP) is a standard mode of operating throughout the marine industry; without it much of what is commonplace now would be difficult or even not attempted. The Nautical Institute oversees the training and standards of dynamic positioning officers (DPOs). The Marine Safety Group, comprising members of the International Marine Contractor's Association (IMCA), monitors operating standards for dynamic positioning vessels.

By integrating precise satellite or other positioning systems and vessel propulsion, dynamic positioning systems enable vessels to maintain position and/or track, to carry out precise subsea operations. The system requires inputs from the navigation system, anemometers, motion sensors and doppler logs to control the propulsion system. Before undertaking a DP check, the survey team will take a sound velocity measurement and will input the result into the DP system.

Loss of DP control is an exceptionally rare event. If it occurs, it is a serious matter that requires thorough investigation and quite possibly a length of time off task to identify and rectify any defects. An extensive field entry DP trial is required to check and to document that the DP system is functioning correctly. If the vessel is to enter a 500 m safety zone or is working under a permit to work, the trial results and a 500 m safety zone pre-entry checklist must be forwarded to the area authority as a precondition of the permit issue.

If a rock installation vessel is working within a 500 m safety zone, it must be operating in DP2 and in a blow-off situation, where the ship would drift clear of assets in the event of a power loss. DP2 requires a second independent positioning reference system to be online, with a third available, and also a certificated engineer to be available in the machinery control room. In addition the ship must be operating to be able to identify worst-case failure limits in existing environmental conditions. In general the ship should use no more than 45% power, which is to ensure a redundancy of power is maintained. Some operators classify DGPS as a single system, irrespective of how many separates systems are available.

The vessel will have to complete a DP check list and submit it to the permit issuing unit for compliance before a permit can be issued. Independent DP trials are conducted annually under the supervision of a classification society and typically take around 24 hours. These trials have been controversial in the past where defects have arisen as a consequence of the trial's more extreme conditions.

Navigation

Vessel navigation is normally understood to refer to the officer of the watch monitoring the vessel's position and progress whilst in transit. The navigation systems in general use today are satellite based, in the same way that survey positioning systems operate. In the recent past, vessel navigation was based on layers from visual bearings of charted features close to land, radar ranges further out, then electronic systems like Main Chain Decca and LORAN C.

Astronomical navigation using a sextant and star tables was the standard method of navigation to monitor position during ocean transits. Astronomical navigation depended upon a clear sky and a discernible horizon. Its prime use was to monitor progress and make a landfall when coastal navigation methods could be used.

By the early 1990s that had all changed with the widespread availability of GPS which provided accurate positioning within satellite constellation coverage. In coastal waters some adjustment had to be made to correct the satellite position to the local chart.

Positioning check

A positioning check, sometimes referred to as a gross error check, is a basic pre-work check to ensure that surface and acoustic positioning is functioning correctly. It is particularly important if the new site is at a significantly different water depth from the previous project and if the last ultra-short baseline (USBL) calibration was carried out in shallower water. The first stage of a position check is to calculate the speed of sound in water at that location using a probe which will measure water temperature, salinity, density and conductivity against depth. The data is extracted and an as-found sound velocity is computed. The sound velocity value is input into the high precision acoustic positioning/hydroacoustic positioning reference (HiPAP/HPR) system before any further checks take place

To eliminate heading errors, the position check should be carried over four passes with the ship's head on reciprocal headings, the main cardinal points tend to be favoured. If the acoustic positioning is not performing satisfactorily the first option is to change the acoustic beacons. If a beacon change does not work the second, though less desirable, option is to carry out an HiPAP/HPR system calibration, which can take around three hours.

The position check should use a well defined and easily identifiable known point such as a well head or structure. For this reason rock berms are unsuitable targets. If a known point is not available an acoustic beacon can be placed on the seabed and position data obtained. The author would advocate using the MBES to detect the beacon and evaluate its position from a series of MBES passes. With a modern system, positioning should be within 0.5% of the slant range from the transducer.

Older structures

The ships' survey and project team should be fully aware of the age of structures in a field. Advances in technology mean that older structures could not have been as accurately positioned as more recent ones. In the last 10 to 15 years there has been a number of significant improvements in GNSS, acoustic positioning and software. When combined, these improvements have given rise to greatly improved accuracy. Older structures which predate the widespread introduction of DGPS or which were installed in its earlier years may be accurate to a number of metres, which can be problematical in itself as explained below.

Modern acoustic navigation systems combine a GNSS derived position with an acoustic position derived from the HPR. It should be accurate to within 0.5% of the slant range. For most practical purposes a rock installation vessel HiPAP/HPR system will be operating within the limits of the scope of movement for the fall pipe. It will not be operated at significant offsets from the vessel, thus reducing some source of error induced by operating at the extremities of the higher angular offsets.

Older infrastructure positions could be metres out from one derived using modern positioning systems. Modern GNSS systems can be accurate to around one metre; this is a big improvement when compared with an accuracy of four to five metres which was the best that could realistically be achieved with the first dGPS. When combined with earlier acoustic navigation systems a position within 1.5% of slant range could be obtained.

If there is a difference between the as-given position of an older structure and a series of as-found position checks, the as-found position check data should be examined for consistency. If the as-found positions are consistent then it is advisable to consider obtaining more recent position information to confirm the as-found position. If data obtained by modern standards is not available, a consistent set of results from the position check should provide confidence that the positioning system is functioning properly, even if it there is a discrepancy in position by some metres. In a rock installation vessel, the position check and the MBES check can and probably should be combined.

The author has recently witnessed a consistent position check on an older structure where all four as-found observed positions agreed to within 0.1 m. These four positions were three to four metres different from the as-given position on the field data. The author had knowledge of the site from a previous contract and was satisfied that the vessel positioning was correct. He obtained recent inspection data from the project staff to confirm that the as-found observed position agreed with other recent as-found data. The contractor was able to start works confident that the positioning was functioning correctly.

Acoustic beacons

The deployment and recovery of beacons without a WROV is difficult. However, despite being rather cumbersome, some fall pipe ROVs are fitted with a hydraulic arm which can be used to deploy and recover a beacon, which must be fitted with a large semi-rigid recovery strop for ease of recovery. Deployment is straightforward, but recovery can be a long-winded process. One company only uses an acoustic release and recovers the beacon with the ship's fast rescue craft (FRC). In unsuitable weather conditions the beacon is abandoned on the seabed, which is wholly undesirable for a range of reasons.

In a project where a number of vessels will be deployed, the project surveyor or delegated offshore senior surveyor will coordinate beacon codes and their management between vessels. This is to avoid mutual interference in the field. The lead vessel senior surveyor will publish a list of beacon codes in use.

MBES calibration

A rock installation vessel carries two or three multi beam echo sounders, two of which are usually placed on the fall pipe ROV and one may be hull mounted in the moon pool area. MBES is corrected for yaw, pitch and roll and needs inputs for the velocity of sound in water. A hull mounted MBES is very useful when working close to platforms and structures, especially in shallower water. Survey teams will be aware of the coverage available from MBES which at a given depth is usually 3–4.5 times the height of the transducer from the seabed. An MBES operating close to the seabed has the benefit of using the highest frequencies at short transmission pulse lengths to give the fullest resolution whereby swathe width of coverage is sacrificed for high resolution. Swathe width can be extended to enable a single pass if two MBES transducers are set up at an angle in the profiling mode. The survey team can compute a height above the seabed to enable the desired width of cover to be achieved, although in a wide corridor or in localised areas, additional lines may be required.

Pre-installation surveys

After the position check and any calibrations have been carried out, the contractor will always carry out a pre-installation survey. Despite the improving accuracies of acoustic positioning it is essential that the target is positively identified if it is on the seabed or in an exposed trench. Rock installation is to some extent based upon relative positioning, as parameters such as depth of cover and top plane are always specified in terms relative to the pipeline, cable or structure of interest.

The contractor also needs to know the seabed topography and any free span of pipe or cable in order to adjust the designs to the as-found survey and thus plan the rock installation. Most rock berms are designed to provide from 0.7 m to 1.0 m of rock protection in all directions from the point of interest. The pre-installation survey data is gathered using the fall pipe ROV mounted MBES, altimeter and bathymetric depth sensor. Pre-installation survey processing can be carried out whilst the fall pipe is being built up and the ship is closing the start point.

A pipe or cable on the seabed is readily detectable with MBES systems. Buried lines cannot be detected and therefore any indication of position such as a trench or spoil can be used to refine the pipe or cable position. In a soft seabed there may be no detectable evidence of the pipe or cable and third party as-trenched or as-ploughed positions may have to be used. The use of third party data without a position check using the third party's data as a control should be avoided if at all possible.

The superintendent may plan to carry out a series of pre-installation surveys to minimise the number of non-productive fall pipe movements and transits. The plan will usually be to start surveys to evaluate the first of the berms to be built. Checking and approval of the design profiles can then be carried out in good time before the vessel is in position to begin to carry out any rock installation and without the anxieties of the vessel being on standby whilst survey data is being finalised. The berm design profile is superimposed upon the survey results as a series of cross profiles normally at 2.5 m to 5 m intervals and a longitudinal profile taken along the centre line. The given design data is then presented to the client for checking and approval.

Fall pipe build up point

Most client companies will not allow any fall pipe rigging or unrigging operations to take place near subsea assets. To mitigate the risks of dropped object damage when handling the fall pipe sections, the fall pipe build up point is set to be a safe distance, clear of any structure. This distance depends upon the operator; some specify a fixed distance, whilst others require the fall pipe build and recovery point to be a multiple of the water depth. The route to and from a proposed build up point should be surveyed to determine a safe route to the start point. This simple precaution should eliminate the risk of grounding the fall pipe ROV on unsurveyed seabed features or, worse still, charted seabed structures where their height above the seabed has not been measured or taken into account.

The start point

The rock installation team will calculate the time needed for rock to descend the length of the fall pipe and then onto the seabed. When the fall pipe is built up and the survey is processed the start position will have been calculated. The length of the fall pipe is known and the rock descent velocity is known. The start point is calculated so that the descending rock exits the fall pipe and reaches the seabed at the start of the planned rock berm. A stop rock position is calculated to keep rock within the berm area as the fall pipe reaches the end of a run.

Method and monitoring

The superintendent will work out how the rock berm is to be laid. The following needs to be taken into consideration:

- seabed topography
- the berm design
- tolerance specifications
- the size of rock in use
- any other client requirements.

A long berm is worked in suitably sized sections to enable the survey team to keep up with post-SRI survey data and to reduce the time between completing the last survey and the data being available for client review. The SRI run is designed to start at the deepest points in an undulating seabed and parallel to the berm long axis in a series of runs, sometimes referred to as strings. Unless the vessel is working in shallow water, there will usually be some scope to manoeuvre the fall pipe.

Some berms, such as those built to support jackup rig legs will have a very tight vertical tolerance. Rock is placed to preserve the berm's stability and to achieve the tolerance efficiently. Rock installation contractors will develop their own methods based on their equipment and experience to install these and other custom structures to the required specification.

The rock installation companies are always looking at efficiency of operation. In recent years more modern vessels have faster fall pipe deployment and recovery times. Time spent loading can be reduced depending upon how well the vessel matches the quarry's loading system.

Operating the fall pipe

The fall pipe can be configured for a task and most companies have a computer program to enable the optimum settings to be obtained for a particular task.

The superintendent, deck engineers and fall pipe operators will calculate the number of fall pipe sections needed and their order in the fall pipe. The telescopic section is uppermost in the fall pipe and it is also secured by stabilisation wires to hold the fall pipe in place laterally when in use. The telescopic section provides vertical movement of 6 m to 10 m depending upon the vessel. A water inlet pipe is placed below the telescopic section and then the rest of the pipe is made up from a combination of heavy duty and standard sections. Without a water flow, the descending rock forces water out from the bottom of the pipe, at up to 30 metres below the actual water level. This creates a pressure differential with low pressure in the pipe and this can cause the upper fall pipe sections to implode.

The free flow of water keeps the rock lubricated and allows the density of material in the fall pipe to be maintained at a desired 1 Te per cubic metre and to maintain an exit velocity that prevents the flow of rock from acting as a jetting tool and displacing seabed or rock berm material. The free flood holes in some systems can be adjusted as required for the operation. Too much water can increase the rock exit velocity to unacceptable levels while too little can lead to a reduction of pressure inside the fall pipe and risk its collapse.

Each fall pipe type has its pros and cons and in the author's opinion neither provides a significant competitive advantage. In deeper water there is an advantage to a speedier deployment or recovery. A time saving of two or three minutes per fall pipe section when working in deep water can save many hours or even days over the course of a long project. The fall pipe can be set for a particular task. Each company has developed tools to enable the superintendent to manage the fall pipe and the rock installation process.

To direct rock flow more accurately, each company has developed fittings which are attached or incorporated to the base of the fall pipe ROV. These can be controlled by the rock installation team to direct or deflect the flow of rock towards, or away from an area of interest as required. Controllable flow devices are very useful when working close to structures or placing rock to secure a PLET cover or to slow the rock flow speed.

Rock placement

The offshore manager will produce a dump plan which summarises the scope of the work and, each berm will be specified to cover position, size and quantity of rock required. The shift personnel will place rock in accordance with the dump plan. The rock installation team can control the volume of deposition at a location by using a combination of vessel speed, conveyor belt speed and the rate of release from the hoppers.

Most companies use dump tool software to assist with the rock placement detailed in the dump plan.

Each company has corporate knowledge of the most efficient way of placing rock and superintendents know their systems very well. It is usually desirable to have berms at stages of completion to keep rock flowing whilst survey data is being processed and the results evaluated.

Intermediate surveys

The rock installation team will monitor the multibeam as a downward looking sonar and in the early stages will be able to detect the pipeline or cable if exposed. When a section has been completed, a survey is carried out to check for high spots or depressions in the berm. The rock installers have a display of the latest DTM, using colour banding to highlight the areas in need of attention. This is used to direct the rock installation team to the work area to bring the rock berm to specification. If the survey shows that the rock berm needs further work, it is referred to as an intermediate survey and is generally a tool for the contractor's use. The contractor should usually only present surveys of those areas that are in specification for approval by the onboard client representative.

Post-survey

When the superintendent is satisfied that the rock berm meets the client specification, it is referred to as post. The post-survey results are prepared for presentation to the client for approval. The pre-installation and the post-installation surveys are the source data for charting as part of the deliverables.

Simultaneous operations (SIMOPS)

In many projects a number of vessels may be working in the same field at the same time. The project team may manage these vessels to keep them apart within acceptable distances. Some activity may be incompatible so the project team ashore will have to allocate priorities to vessels to complete their tasks.

Close approaches and SIMOPS

Close approaches to a platform

Close approaches to a semi submersible drilling rig within its anchor pattern

SIMOPS can be managed at the daily conference call or at a separate meeting. Rock installation activities can degrade visibility so it is often good practice to keep SRI vessels in areas well away from the site of operations of other vessels. The author is of the opinion that some degraded visibility is also attributable to ROV and diver activity in a soft seabed of fine material.

In a mature field there will be platform service vessel (PSV) movements to and from platforms or FPSOs. Operator company marine operations procedures may restrict the number and activity of vessels within their 500 metre zones. For example, a company may prohibit simultaneous rock installation and PSV operations on the same rig face within a 500m safety zone.

The field operator will provide a marine operations manual for the guidance of marine and project crews. Other naval, commercial and recreational shipping will be seen in transit through oil and gas fields.

13 The final product

The client will specify what deliverables are required following rock installation and how it is to be presented. Reports and graphics are required and electronic copies can be provided as an alternative to paper copy. Most reports will form listings and technical data, although some narrative may be necessary. In general, it is desirable to explain what was done and how. It should not be forgotten that it is of equal importance to give an explanation of any shortcomings and the reasons for them.

Deliverables

The contract deliverables are reports and graphics as a series of continuation sheets to cover the rock berm, in either A0 or A1 paper size. Graphics are increasingly being delivered electronically as pdf files and in other formats as required by the client company. The continuation sheet will have a title panel, usually on the right, showing the project details, geodetic parameters, sheet details and both horizontal and vertical scales. The first and second data panels will contain a plan view of the pre- and post-installation bathymetry showing an outline of the berm design in both cases. The plan views will display contour lines which help to give a depth of field, kilometre post (KP) values and the pipeline route, which becomes the long profile. Cross profile locations are shown perpendicular to the longitudinal profile. A longitudinal profile showing the berm, design and tolerance level is presented in the third panel. Finally, cross profiles with the design and tolerance profile are shown in the fourth panel. These are typically delivered at a five metre interval, although other intervals may be specified.

An example of a final chart is in Appendix G. The pre- and post-lay surveys are presented on a skew with a north arrow to enable orientation.

Reports will comprise a narrative and listings; typical reports will be a survey mobilisation and calibration report to demonstrate that all components in the survey system are in date for testing and functioning correctly, and an as-built report covering all aspects of the installation. Anomalies may be covered in a field report, apart from field calibrations; the author has rarely seen field reports during rock installation tasks. A field report is a useful tool to get information to the users and interested parties expeditiously.

Pre-SRI plan view	Notes
Post-SRI plan view	Legend
Longitudinal profile	Scale Project details
Cross profiles 1 or 2 panels	Logos Signature Block

Panels making up a typical SRI continuation sheet

Summary

A successful rock installation phase starts with the initial desktop study, the project team can use data from some or all of the following sources:

- Admiralty and national nautical charts
- national mapping and geological authorities
- oceanographic data including tidal stream and ocean current information
- company survey data
- ploughing and trenching data in the locality
- third party data, for example other company data and fishermen's charts
- nautical publications containing meteorological statistics
- academic papers

If necessary, approval should be obtained and enquires made of any particular requirements from third parties regarding crossings.

If pre-route and route surveys are scheduled, there is an opportunity to obtain geotechnical data with grab samples, core samples and cone penetration tests.

The purpose and specification of the rock berm must be identified for example:

- protection from impact, trawls or anchors
- scour protection in shallow water with stronger tidal streams
- support over a topographic feature such as deep glaciation scours
- pre-lay berm for a structure or pipeline
- burial of a disused asset

- stabilisation for upheaval and lateral buckling mitigation
- containment

It is always most beneficial to enter a dialogue with the contractor as soon as practicable to discuss:

- the right vessel for the job
- provisional timetable
- an estimate of the quantity of rock required and supplier
- whether the specification and proposed material match
- delivery of pre survey back ground files
- deliverables
- any unknowns such as post out-of-straightness calculation stabilisation berms

As a project is pending, the project team will have to

- apply for any permits such as a DEPCON, which should have plenty of contingency
- agree the specification
- authorise documentation
- prepare a project plan
- carry out any liaison if required with fishermen's organisations

Once mobilised, personnel must be nominated to have responsibility for:

- obtaining and managing permits
- managing any third party requirements and permits; some operators may require representation on board for crossing works, others will not
- managing simultaneous operations with other vessels, platforms, etc.
- reporting progress
- reporting any non-conformance with the contractual specification or any regulation
- enabling day-to-day liaison with the project staff and interested parties
- compliance with any national licensing conditions, e.g. UK DEPCON, regarding berm location, quantity of rock allowed and for monitoring licence time limits

- drafting deliverables including final reports, graphics and listings for project and office QA/QC before delivery

Finally, before departing the site for demobilisation, the data should be checked for completeness and any shortcomings or locations for remedial works should be recorded and reported, especially if the vessel is departing the field empty.

Deliverables will be subject to the contractor's office QA/QC procedures and there may be a gap in time between work finishing and the transmission of deliverables. The project staff will be keen to review them. In the author's opinion draft deliverable data which has not been through the QA/QC in the contractor's office should not usually be passed to the client. It will invariably lead to avoidable paperwork should a reviewer be in possession of two slightly different data sets.

Conclusions

Rock installation has many current practical uses in a broad range of projects and no doubt the future will see more innovative and new applications. The idea is simple; however, in practice a successful operation requires up-to-date background information, a specialist vessel with the facilities to provide complicated, well developed control, ballasting, rock delivery, and the handling systems to enable highly skilled personnel working in an integrated team to manage the rock cargo from loading through to installation.

The ships need to be equipped with precise satellite and acoustic navigation systems and to have highly skilled and committed multidisciplinary crews to complete works, sometimes within very tight technical specifications and also time and budget constraints.

A rock berm can provide an environment to support a full range of marine life in an otherwise sparsely populated seabed locality. As happens with wrecks, a rock berm will support a diverse population of fish and if large enough, will be attractive to commercial fishing interests.

Rock installation is not a cheap procedure; on the other hand the client company can be confident that it will deliver the cover, protection, support or stabilisation needed. Cheapness can be replaced by the added value of overall cost effectiveness.

A comprehensive desk study should form the base for advance planning. This study may determine the extent of any fieldwork required. The results of the study and surveys can be used in a dialogue between the end client, other partners and the contractor to ensure a successful rock placement phase of the project.

Uses for rock installation have expanded as new needs arise from the energy sectors, coastal civil engineering and compliance with regulation. Rock installation is a versatile option on its own or when combined with a trenching and ploughing campaign. The development of shallow draught vessels with a smaller rock installation capability will continue to have an important role in the full range of coastal engineering and shallow water projects.

The future is bright with oil and gas, civil engineering, renewables and power distribution cables all requiring SRI. As older oil and gas assets are decommissioned there may be a case for placing a rock berm over the remains of platform legs and other infrastructure. In some older fields new technology has enabled a longer life and old unserviceable pipelines are being replaced.

In coastal engineering, SRI can have a place in coastal protection, although it is a highly complicated issue and one area's protection may have undesirable effects further along the coastline.

Appendices

A Useful websites
B Marine, personnel and security
C Table of current fall pipe vessels
D Map showing 11 Norwegian coastal quarry locations
E DEPCON calculation step by step guide
F Diagrams showing representative types of rock berms (not to scale)
G Example of a post-rock-installation chart
H Representative rock use
I Typical mission profiles (% time)

Appendix A Useful websites

The following websites may be of use when planning subsea rock installation. This is not an exhaustive list and other unspecified websites may be available that provide useful information. This list is correct at the time of writing (May 2019) and readers are encouraged to check up-to-date sites and information.

Subsea rock installation contractors

Boskalis www.boskalis.com
DEME / Tideway www.deme-group.com or www.deme-group.com/tideway
Jan de Nul Group www.jandenul.com
Van Oord www.vanoord.com

National authorities and government agencies

Danish Energy Agency www.ens.dk
German Government and states www.energytransition.org
Mexican Department of Energy www.gob.mx/sener/en
Netherlands Government www.government.nl/topics/renewable-energy/offshore-wind-energy
Norwegian Geotechnical Institute http://www.ngi.no
Norwegian Oil and Gas Department www.regjeringen.no
Norwegian Petroleum Directorate www.npd.no
UK Oil and Gas Authority www.ogauthority.co.uk
UK Marine Management Organisation www.gov.uk/government/organisations/marine-management-organisation
UK Health and Safety Executive offshore oil and gas www.hse.gov.uk/offshore
UK Health and Safety offshore wind farms www.gov.uk/horizons/assets/documents/wind-energy.pdf
USA North American Offshore Energy www.doi.gov

Trade associations

European Wind Energy Association www.ewea.org
Global Wind Organisation www.globalwindsafety.org
International Association of Dredging Contractors www.iadc-dredging.com
International Marine Contractors Association www.imca-int.com
Norwegian Mining and Quarrying Industries Association https://www.norskbergindustri.no
Oil and Gas UK www.oilandgas.co.uk
Renewable UK www.renewableuk.com

Professional bodies

The Chartered Institution of Civil Engineering Surveyors www.cices.org
The Institute of Marine Engineering Science and Technology www.imarest.org
The Royal Institution of Chartered Surveyors www.rics.org
Society of Underwater Technology www.sut.org
Subsea Engineering Society www.subseaeng.org

Appendix B Marine, personnel and security

ISPS

The International Shipping and Port Facility (ISPS) regulations cover ship and port security. All personnel movements, including visitors, have to be notified to the port security managers via the ship's agent in good time. Many ports require 24 hours notice for visitor access. Personnel proceeding ashore must carry recognised identity documents and be advised of any codes or other security measure to gain access to the port.

Depending upon the security state in force, vessel access will be restricted and additional security measures may be ordered.

Personnel documentation

International Labour Organisation (ILO) Marine Labour Convention of 2006 regulations have been adopted by many flag states. All personnel working on board may be checked for compliance.

All personnel should have:

- a valid passport
- a valid seamans' discharge book.
- a contract of employment or assignment schedule
- recent pay slips
- international vaccination booklet or certificates if appropriate
- copy of employer's liability insurance certificate
- valid medical certificate e.g. ENG1, Oil and gas UK
- valid safety training certification BOSIET.FOET, HUET, MIST, STCW, or GWO
- evidence of training and competency
- CV and job description
- valid visa or letter of invitation if appropriate

Appendix C Table of current fall pipe vessels

Current fall pipe vessel capacities are tabulated below. The majority of the vessels are relatively new in service. *Sandpiper*, *Rocky Giant* and *Tertnes* have all been retired.

Vessel name (In alphabetical order)	Operator	Nominal rock cargo (tonnes) see notes [1]	Year built (converted) [2]
Adhemar de Saint-Venant[3]	Jan De Nul Group	4,750	2017
Bravenes[4]	Van Oord ACZ	12,000	2016
Daniel Bernouilli[3]	Jan De Nul Group	4,750	2017
Flintstone	Tideway (DEME)	18,000	2011
Isaac Newton[3]	Jan De Nul Group	12,500	2015
Joseph Plateau[5]	Jan De Nul Group	31,500	2013
Livingstone[3]	Tideway (DEME)	12,500	2017
Nordnes	Van Oord ACZ	24,000	2001
Rockpiper	Boskalis	24,000	2011
Seahorse	Boskalis & Tideway Joint Venture (JV)	17,500	1983 (1999)
Simon Stevin[5]	Jan De Nul Group	31,500	2011
Stornes	Van Oord ACZ	26,000	2012
Tideway Rollingstone	Tideway (DEME)	11,000	1979 (1993)

Notes

1. Information obtained from company websites. The actual capacity may be reduced or increased slightly depending upon factors such as fuel load, additional rock handling equipment fitted and time of year.
2. Full details of the vessels are available on operator's websites.
3. These vessels are described as multipurpose vessels, with normal SRI capacities and a large diameter inclined fall pipe for the installation of large size rocks (up to 200 kg). Jan De Nul Group vessels have a trenching capability and are fitted with an active heave compensated crane for

mattress laying and light construction tasks (with the offshore renewable industry in mind). *Isaac Newton* is currently used as a cable lay vessel.

4. *Bravenes* is an X bow™ multipurpose vessel.
5 *Simon Stevin* and *Joseph Plateau* are sister ships, the other vessels are custom designs or conversions.

Appendix D Map showing 11 Norwegian coastal quarry locations

Locations of typical quarries used frequently by the offshore industry		
1. Lervik near Oslo	5. Skipavika near Bergen	9. Nord Fosen near Trondheim
2. Dirdal near Stavanger	6. Bremanger near Florø	10. Aquarock near Sandnessjoen (closed 2020)
3. Jelsa near Stavanger	7. Visnes near Molde or Kristiansund	11. Hammerfest
4. Sløvag near Bergen	8. Averøy near Kristiansund	

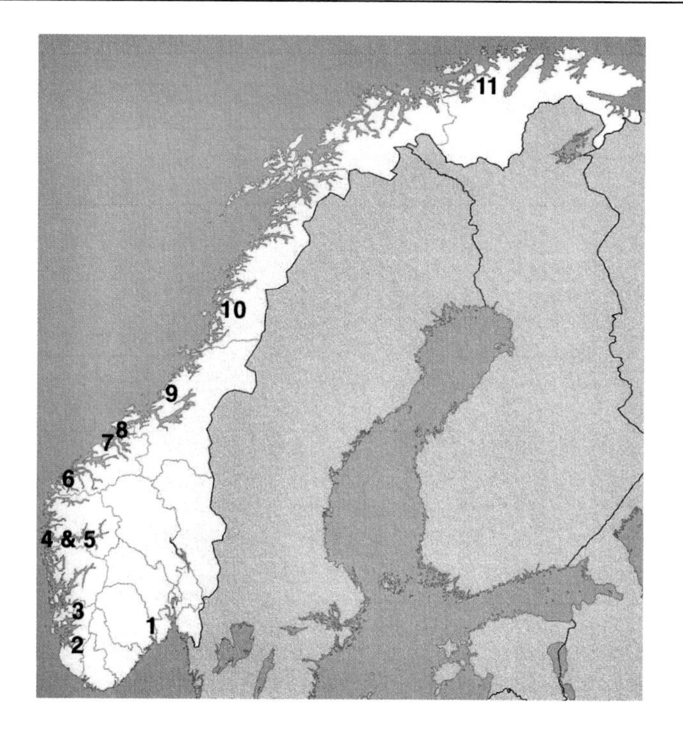

Appendix E DEPCON calculation step by step guide

The operator has to apply for the DEPCON, which is really a permit or envelope in which to operate. The baseline protection workscope is known, however the full extent of any stabilisation or UHB (upheaval buckling) mitigation berms is often not available until work is in progress. The following flow diagram may help to apply for a DEPCON, in which to operate with confidence that the quantity of rock specified will not be exceeded. The contractor will have greater flexibility if the DEPCON covers a whole field or pipeline to include spools, crossings, and covers, etc.

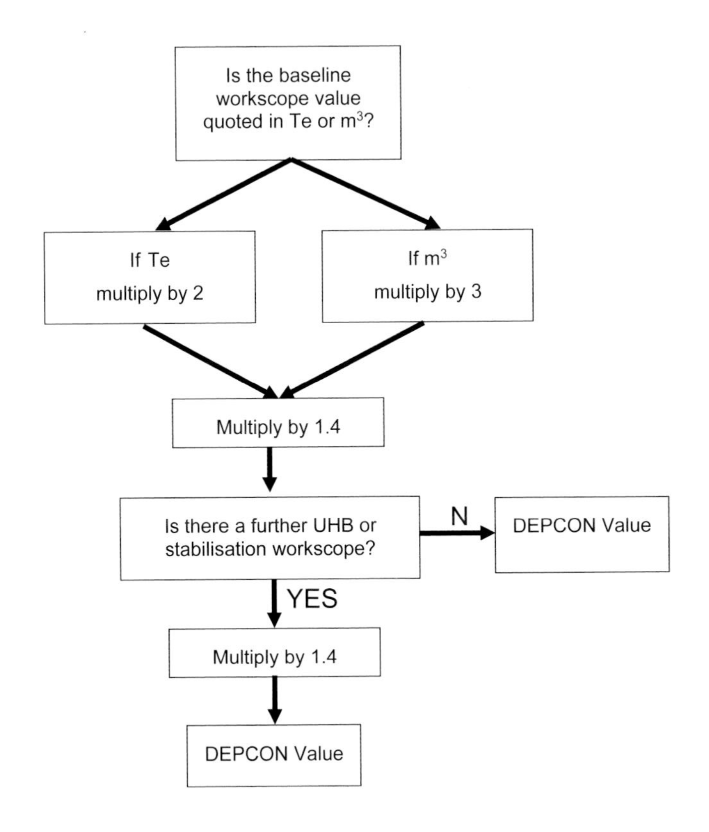

Appendix F Diagrams showing representative types of rock berms (not to scale)

Typical pipeline, flexible or cable berms (Representative and not to scale)

PLAN VIEW

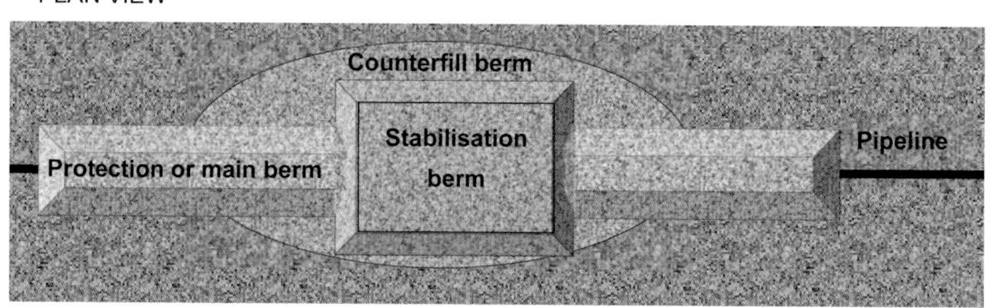

Stabilisation berm

Protection or main berm

Pipeline

Counterfill berm

CROSS PROFILE

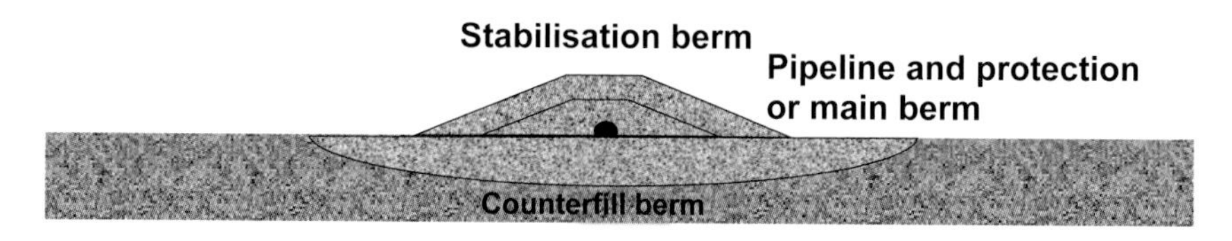

Supporting berm

LONGITUDINAL PROFILE

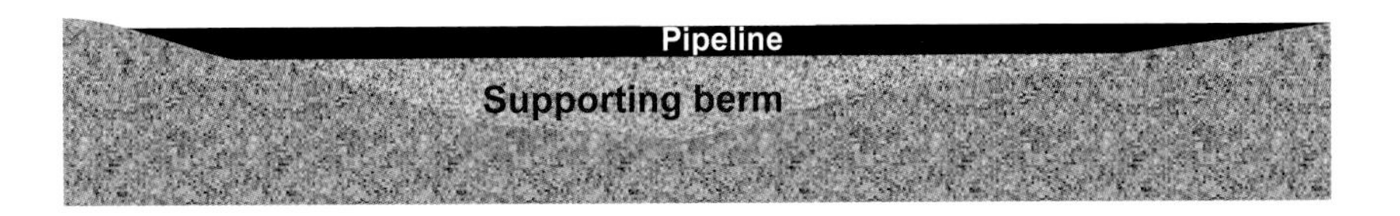

PLAN VIEW

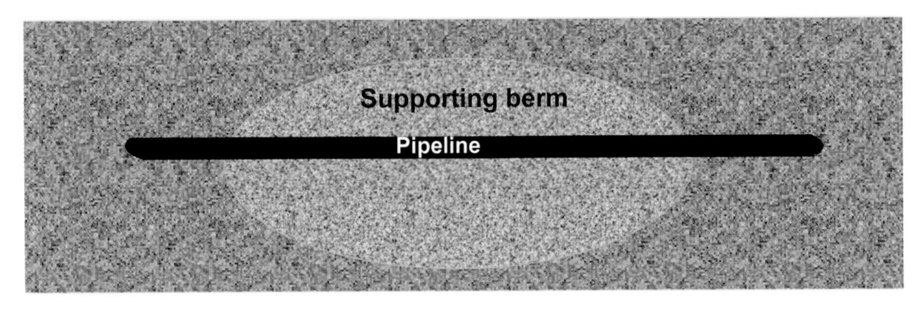

CROSS SECTION

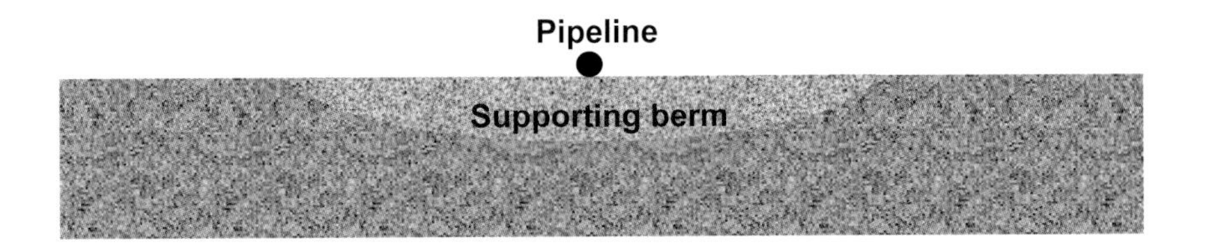

Wind turbine tower anti-scour berm

SECTION VIEW

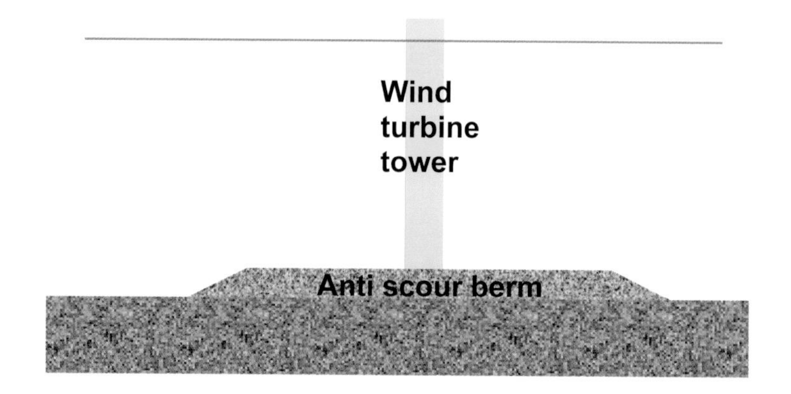

PLAN VIEW

Supporting pads for a jackup rig

Section View

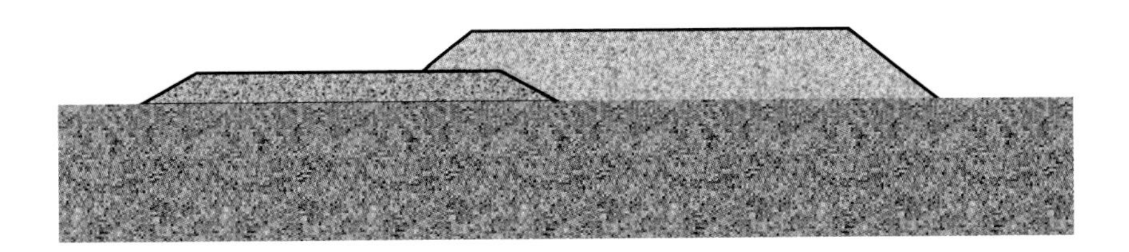

Plan View

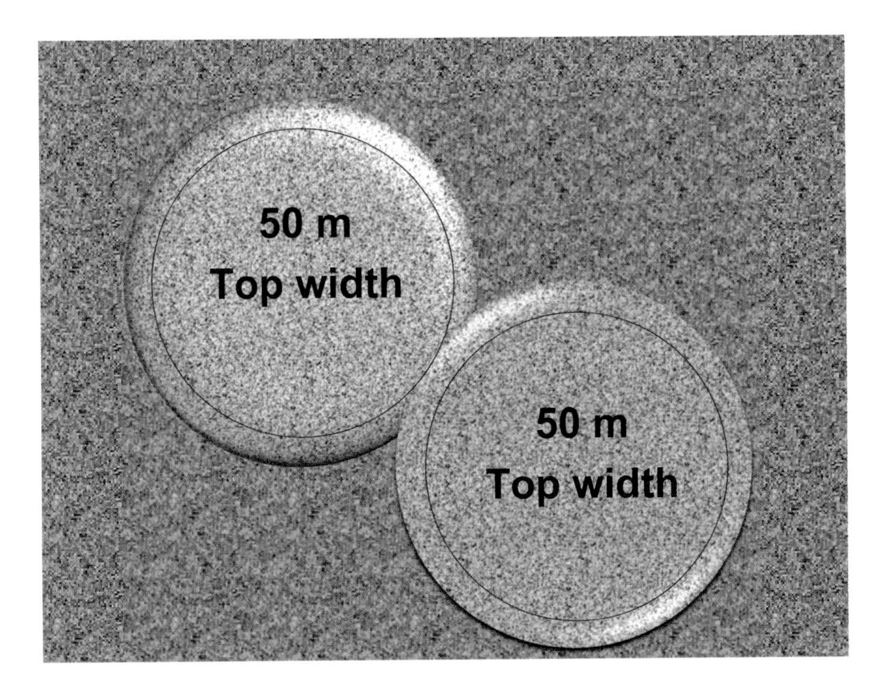

Crossing berm

Long Section View

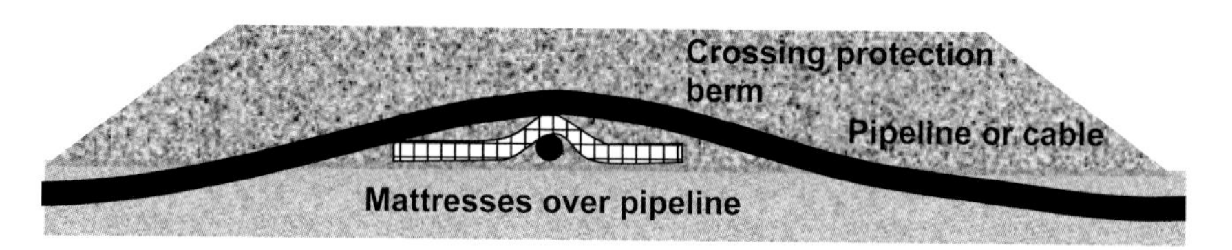

Plan View

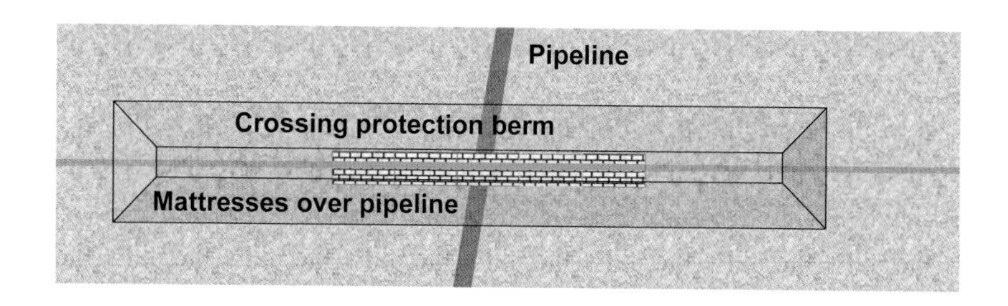

PLET cover protection

Plan View

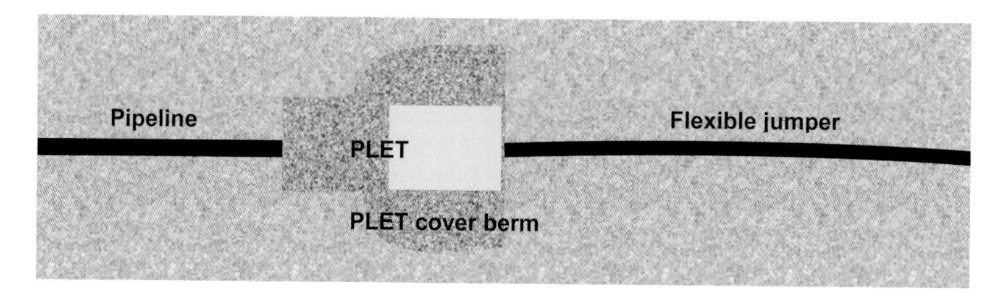

Longitudinal Section View

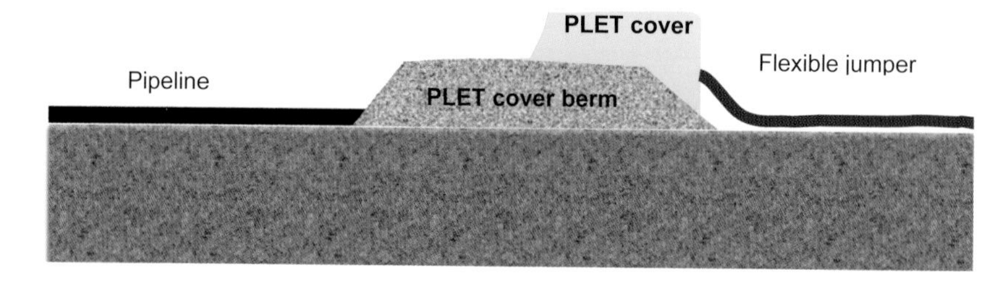

Mattress protection berm

Plan View

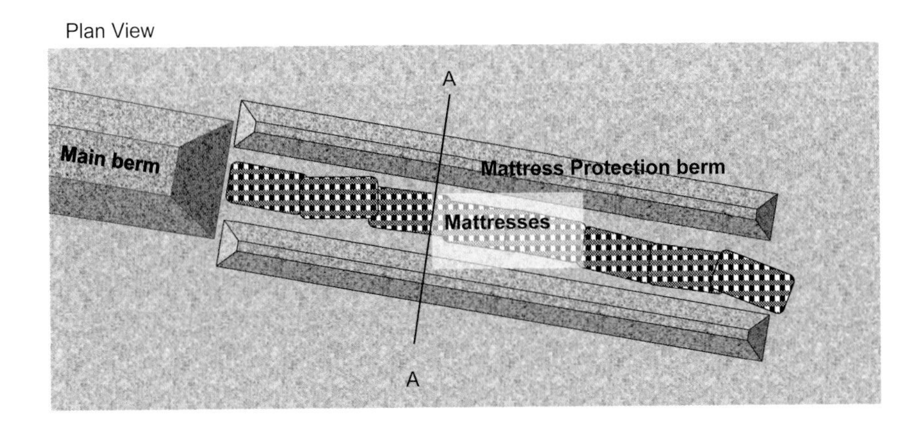

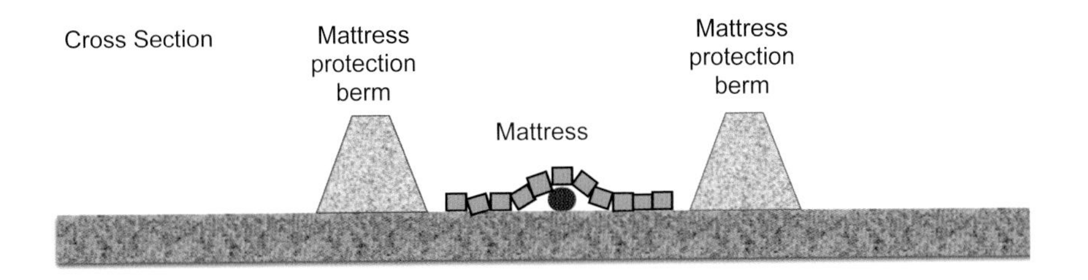

Coverage based on a tonnes per metre value in a well-formed trench

Cross Section of Well Formed Trench in Stiff Material

Armour rock and filter or cushion rock layers

Cross Section Showing Filter or Cushion Rock and Armour Rock Layers

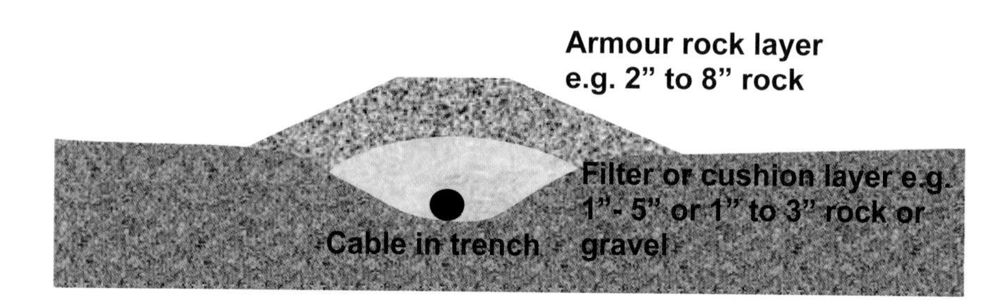

Appendix G Example of a post rock installation chart

(© Enquest Britain Limited and reproduced with their kind permission)

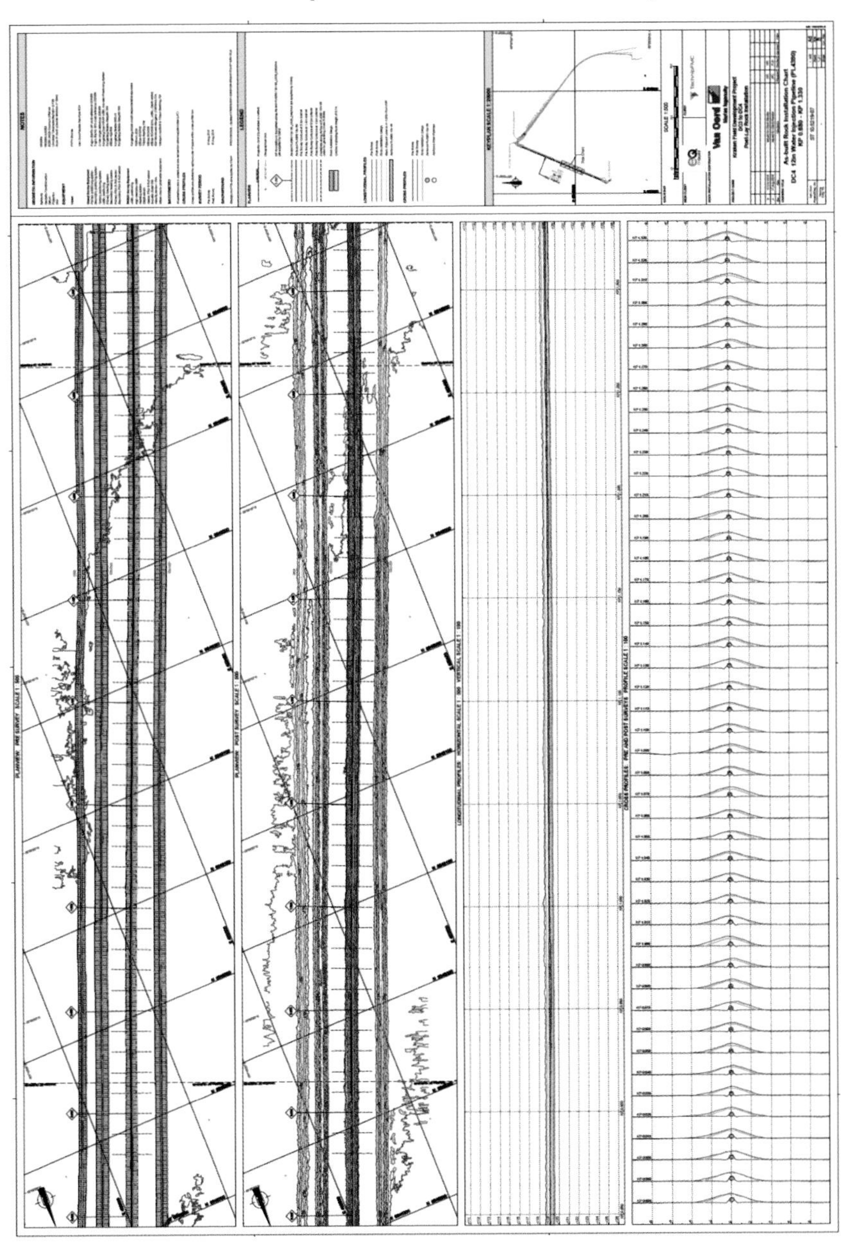

Section of chart showing detail

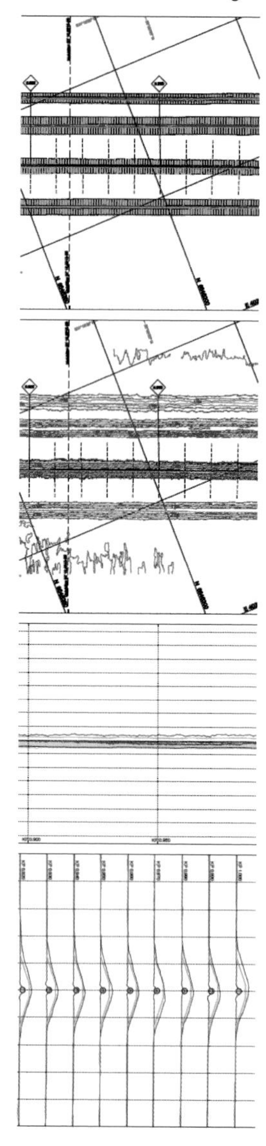

Pre-lay survey showing the route in blue, berm footprint in green and cross profile sections

Bathymetric contour at 0.2 m intervals

Post-lay survey showing contours of the berm at 0.2 m intervals and cross profile sections shown further below. The contours are close together representing the gradient of the berm in an otherwise flat seabed

Longitudinal profile of the centre line showing the top of pipe in blue, the design berm in green and the as laid berm in red.

KP value

KP value

Cross profiles at 10 m intervals showing the seabed and top of pipe in blue, the design profile in green and the as laid profile in red. A black circle represents the pipe.

Appendix H Representative rock use

The tables below shows a representative quantity of rock for products with an outer diameter of 300 mm in the upper table, and 150 mm (typical umbilical) with standard side slope of 1:3, a vertical tolerance of 0.0 m plus 0.4 m. The calculations in the table are rounded to the nearest 100 Te and assume a flat seabed using a conversion factor of 2.2 Te per m^3. No allowance has been made for the volume of the product

130 mm diameter product

Top width (m)	Ht above Product (m)	Lower berm vol m³ per m	Upper berm vol m³ per m	Lower Te/m	Upper Te/m	Lower Te/km	Upper Te/km
0.5	0.6	2.9	5.7	6.4	12.6	6,800	12,600
0.5	1	5.7	9.5	9.5	12.1	12,600	21,000
0.5	1.5	10.6	15.6	23.4	34.4	23,400	34,400
1	0.6	3.3	6.4	7.4	14.0	7,400	14,000
1	1	6.4	10.4	14.0	22.8	14,000	22,800
1	1.5	11.5	16.7	25.4	36.8	25,400	36,800
3	0.6	5.1	9.0	11.7	17.0	11,300	19,800
3	1	9.0	13.8	19.8	30.3	19,800	30,300
3	1.5	15.1	21.1	33.3	46.5	33,300	46,500

150 mm diameter product

Top width (m)	Ht above Product (m)	Lower berm vol m³ per m	Upper berm vol m³ per m	Lower Te/m	Upper Te/m	Lower Te/km	Upper Te/km
0.5	0.6	2.1	4.5	4.5	10	4,500	10,000
0.5	1	4.5	8.0	10	17.6	10,000	17,600
0.5	1.5	9.0	13.6	19.8	30.0	19,800	30,000
1	0.6	2.4	5.1	5.4	11.3	5,400	11,300
1	1	5.1	8.8	11.3	19.3	11,300	19,300
1	1.5	9.8	14.7	21.6	32.3	21,600	32,300
3	0.6	3.9	7.4	8.7	16.4	8,700	16,400
3	1	7.4	11.9	16.4	26.2	16,400	26,000
3	1.5	13.1	18.8	28.9	41.3	29,000	41,300

Appendix I Typical mission profiles (% time)

Multiple load missions were between 5 and 8 loads per vessel. All missions were in the North Sea in different vessels from 2009 to 2019 (values are rounded to add up to 100%)

Activity	Mission 1	Mission 2	Mission 3	Mission 4	Mission 5	Mission 6	Mission 7
Loads	multiple	multiple	multiple	multiple	multiple	partial	partial
Season	winter	winter into spring	spring	summer	summer	autumn	spring
Mobilisation and testing [1]	5.6%	4.0%	1.9%	0%	0%	0%	0%
Transit	28.3%	23.8%	25.8%	24.0%	29.2%	40.0%	19.0%
DP checks	0.2%	0.3%	0.4%	2.0%	0.3%	4.0%	3.0%
In-field transit	7.9%	1.8%	4.0%	4.0%	2.8%	6.0%	2.5%
Tracking to safety zone [2]	NA	NA	8.5%	NA	NA	NA	2.5%
Loading rock	8.0%	5.7%	8.8%	13.0%	12.2%	14.0%	NA
Fall pipe handling	4.9%	3.3%	2.9%	2.0%	4.5%	4.0%	6.7%
SRI[3]	24.5%	15.7%	15.3%	20.0%	37.6%	21.0%	10.3%
Survey and monitoring	6.4%	2.8%	9.2%	6.0%	8.5%	11.0%	9.0%
Weather standby [4 5]	9.2%	39.5%	17.4%	0%	0%	0%	0%
Client standby[5]	1.2%	0%	0.8%	0%	0.1%	0%	0%
Quarry standby[5]	0.7%	0%	0.7%	0%	0.4%	0%	0%
Bunkering & logistics[1]	0.9%	0.9%	0.3%	0%	0%	0%	0%
Breakdown[5]	1.2%	0.7%	1.0%	1.0%	1.8%	0%	9.0%
Other (inc. third party works)	1.0%	1.5%	3.0%	28.0%	2.6%	0%	38.0%
Total	**100%**	**100%**	**100%**	**100%**	**100%**	**100%**	**100%**

Notes

(This table is representative only and accommodates individual contractor's terminology)

1. May be coincident with other activity such as loading.
2. Surveyed area used for transits between sites.
3. Excludes any other activity.
4. A vessel's operating envelope varies with activity and DP system performance.
5. A single short delay can have a big effect upon a short duration mission.

Index